KINEMATICS
of The Brain Activities

MOSTAFA M. DINI PEng

ISBN: 1470082594
ISBN 13: 9781470082598

Table of Contents

CHAPTER III
(claims and their approval)

Preface

This volume is the third volume of a series on the Kinematics of the Brain Activities. The subject is developed in order to make possible an animation model which is capable of simulating a built model based on the information detailed in the book series. The simulated model can demonstrate the predicted outputs which help to modify the model aiming to provide actual behaviors of the brain. A fourth volume which I hope to publish soon will develop the subject further toward the same goal.

In the previous volumes, the chapters were organized according to the brain structural levels: molecular, substrate, pathway and layer. In this volume I have tried to follow the same procedure, although was not possible for some of sections. A comparison between present theories and the Kinematic Model is given in the following to make the main body of the book easier to grasp.

Due to the application of the terms borrowed from fluid dynamics and other fields of physics, some multidisciplinary terms have been used here. At the same time, building a model on a macro scale requires simplification. Therefore, some new terms have been used. For example, the term mentation refers to different currents of feelings, automatic thoughts and intentional thinking, which are attached to, or named for, different types of energy transfers.

Furthermore, some abstract and physical meanings have been assumed to be equivalent. For example, substance names, events and functions are referred to as creeping (substantial or static), transmitting (flowing) or procedural energy flows when a relevant pathway structure is excited.

Introduction

The development of synchronized firings in the brain on a mass scale through a pathway makes possible a kinematic study of brain activities. Kinematics is a branch of physics that perceives the brain as layers through which free energy flows. Therefore, instead of considering the individual brain neuron firings, Kinematics considers a flow of synchronized firings, in packets and in definite routes, and their impacts on the routes. The routes are referred as pathways here.

The development of firings routes can be followed by studying a variety of chemical, electrical, magnetic field energies and tiny changes in blood flow to parts of the brain. However, if all of these energies are measured in relative to any change in firing gates spatial configuration, they become much simpler to study. Thus, there are continuous firing gates configuration changes occurring in excited pathways over the outer layer of the brain in the shape of steady fluctuations. The elasticity of the outer layer remains linear for these steady oscillations. However, when oscillation exceeds for beyond the linear range of elasticity behavior, the location of the non-linear oscillation undergoes strains. The excess of strain related to the overwhelming of the linear limit of layer elasticity is consumed by chemical changes in the place causing other changes in electrical parameter and magnetic field strength. These changes occur in the space of firing gate channels and accelerate the firings that occur there. The confined energy accumulated in the pathways over the layer cause decay in the layer's elasticity property, which is then recovered during sleep in a regenerative process.

Before starting the kinematics of the brain activities, following a brief description of difference between kinematics and neuroscience common terms are given as follows:

Kinematics and Neuroscience

Energy	
Neuroscience	The Kinematic Model
• Considers linear transfer of energy in circuits	• Considers layers of orbits and fractals in complex pathways In parallel and in pulse sequential
• Transfers are in sequence with tiny time lags	• Considers deflected fractals flow
• Considers strings of input-output circuits	• Considers energy layers' configuration, in addition to electro-chemical reactions
• Considers electro-chemical energy types	

Brain structure	
Neuroscience	The Kinematic Model
• Structure is considered static (except in the case of injury)	• Structure can be tensed, releasing or in rest and has the potential to deform
• All phenomena occurs, fundamentally, at the molecular level	• Phenomena occur on the different scales: molecular, substrates and the pathways in layers
• Outer layer surface and folds are constant after growth ages	• Slight changes occur after growth age (possible increase in fine folds), over the course of decades (permanent curvatures) and daily (changes in a few spread curls)

Regeneration	
Neuroscience	The Kinematic Model
• Not in the equation	• Required periodically to retain the previous day's viscoelastic properties
• REM and NREM are described as periodic changes in oscillation frequency	• NREM is needed for new curl consolidation in different levels and REM is used for configuration equalizing (same positions as previous day in any level)

Functions and memories	
Neuroscience	The Kinematic Model
• Static memories are localized in specific locations in the brain; but recently believed that memories and functions can be saved in spread locations	• Memories will be saved randomly everywhere, but they will be saved family-wise (resulting in habits and moods), semi-localized and generally localized. Such a migration in macro-structure happens due to resonance and energy transfer via the self-repetition of an attractor.
	• Or, in patterns; this considers the brain as a chaotic media for spatial energy layers during a mentation.
	• However, the type of structure and the attached energy configurations are closely related.

Oscillation Records	
Neuroscience	The Kinematic Model
• Frequencies are considered as related to synchronized firings.	• Frequencies are considered as related to synchronized firings, em-fields and energy particle variations.
	• Frequency bands are related to the structure of the location.

Brain Activities Explanations	
Neuroscience	The Kinematic Model
▪ No clear explanation for the necessity of sleep, or the reason for and nature of dreaming and memory consolidation	• Comprehensive integrity between different aspects of the model, including the necessity of sleep and the reason for and nature of dreaming and memory consolidation
• There are sequential happenings in locations and certain time lags between activities.	• Parallel occurrence of intermediate activities and a time lag between input and output occurrences.
• Once-through looking to a mentation between inputs and outputs.	• Iterative loops of pathways and intermediate products that complete a mentation relatively by checking the environment for even the smallest possible error.

Local Fiber Structures and Energy Flow Configurations	
Neuroscience	The Kinematic Model
• Brain structure generally remains constant after growth ages.	• There is a two-way interaction between energy flow and brain structure. Stresses induced by energy momentum transfers threaten to deform the structural curvatures while the pathway structure regulates energy layer configurations.
• The subjects of study are molecular components (like genes, DNA,…), ion gates, and neuron cells	• Location structures are categorized for molecules, firing gates, strings of synapses, em-field's radius, substrates, pathways.
• Energy concept is never a primary concern.	• The types of energy assigned to locations are: chemical energy, eddies, energy packets and energy wakes.

Brain as a Chaotic Media	
Neuroscience	The Kinematic Model
• Mostly looking for exact mechanisms in the brain	• Considers the brain a chaotic media for energy configurations in fractal shapes.
• Focuses on deterministic study and rational correlations between molecular phenomena happening in different locations of the brain	• Studies the correlation between location activities from a non-deterministic and chaotic perspective.
• Not physical media explanation for drives and pathway directions	• Defines attractors that determine energy layer configurations spatially and energy transfers as clouds from input back side toward front side of the outer layer.
• Assigns a definite area in stem for brain activities sequence	• Studies the excitation or retrieving of memories in the route of energy layers and resonance phenomena.
• No physical explanation for drives and pathway directions	• The orchestra leader of brain activities is not unique and is led either by input channels in the stem or by internal attractors, changing according to chaos principles.
• No physical explanation for the saving of memories.	• Physical explanation of memory saving as a change in the configuration and size of firing gates and on its extremes molecular changes (like long term adaptations or genetic).

Consciousness	
Neuroscience	The Kinematic Model
• Not a clear differentiation between physics and consciousness in brain.	• Considers consciousness as complex parallel activities confirming or challenging each other in the same activity complex.
• No clear differentiation between unconscious, subconscious, conscious and super-conscious activities	• Considers unconscious, subconscious, conscious and super conscious activities as developed species of fractal complexes through the development of proper structural locations in species' brain.

Brain Global Physical Properties	
Neuroscience	The Kinematic Model
• Viscoelasticity is not a concern	• Viscoelasticity is essential to regulating energy transfers, maintaining all brain properties as a chaotic media, and maintaining memory of past phenomena
• Viscoelasticity is not a concern	• Different viscoelastic properties on different levels: molecular, synapses, fibers, substrates and layers.
• Folds in brain exist to provide more surface area in a limited space of the skull.	• Folds, curvatures and curls are convolutions on different levels that are structured in long or short terms, and permanently or temporarily configure the formation of memories and functions in different locations.

What makes Kinematic Model different from other types of studies?

In what follows, you will find a comparison of the definitions used in present brain activity studies and those which have been introduced in this study:

1) **Scale of units under study:**

- NEUROSCIENCE (Merriam Webster): a branch of the life sciences that deals with the anatomy, physiology, biochemistry, or molecular biology of nerves and nervous tissue and especially with their relation to behavior and learning. At present, Kinematics, in brain studies, is used as a term describing deformations and injuries, normally occurring during accidents or tumor growths.

- Kinematics: a branch of physics that looks at brain as layers that are daily imposed upon with sensory inputs and that undergo some strains; if they are overstrained, this changes the configuration of the firing gates. The brain is a chaotic media on the micro scale level and if firing gates configuration changes, firings in specific pathways are synchronized and form new fractals. Each fractal makes up one energy packet and the growth of new fractals in a sequence and in a pathway is the same as the transfer of energy through that pathway. The brain is a visco-elastic material and energy transfers through its layers follow the related kinematics rules.

2) **Connectivity vs. energy transfer relations:**

- Connectivity (US National Library of Medicine): "the components of a neural circuit are usually distinguished in separate experiments to identify long connections, presynaptic, and postsynaptic components".

- Connectivity (Kinematic): on the macro scale, connectivity is established either between neighboring substrates of energy units or by the resonating of far substrates with multiple frequencies.

3) **Growth of connectivity vs. fiber curvature changes (after growth ages):**

- Neuroscience: connectivity grows through training and adaptation to new environments.

- Kinematics: connectivity grows through daily memory consolidations in sleep—a physical change in fiber configurations. The brain gradually increases in fiber curvatures and in folds and would shrink more during old age.

4) **Sleep nature and necessity:**

- The nature of sleep (Merriam Webster): the natural periodic suspension of consciousness during which the powers of the body are restored.

- The nature of sleep (Kinematics): The brain is strained due to daily sensory inputs and internal activities. The strains accumulated during the day cause the brain layers to lose their elastic properties and to become stiff and inefficient in processing. *Sleep is essential to regenerate and recover the brain's normal properties.*

5) **Dreams:**

- Dream (Merriam Webster): a series of thoughts, images, or emotions occurring during sleep.

- Dream (Kinematics): a series of thoughts and images associated with released energies from the brain's most strained locations; these are released during regeneration. These energy releases are associated with dream images in REM periods and memory consolidations in NREM periods.

6) **Kinematics model:**

- Although it is a combination of unproven theories. However, it is possible to prove those applied theories by looking in contingency in the parts of the integrating theories. A visual project is being defined and the goal is that by providing the software product, a presentation of typical scenarios made of individual memories centered around experienced emotions can demonstrate the model contingency with facts. The software product is based on the assumption that a pathway of memories will be excited in a dream that are attached to the same emotion as has been experienced during a stressful event in the day proceeding the dream. Thus, the success of the software product will practically confirm the correctness of the theories used in the model.

CHAPTER I

Stress Induction, Interaction and Behaviors

Stress flow, straining release or absorb of the strains, accumulation of strain remains during the waking time, decay in elasticity property and regeneration including overstrains release and consolidation are the main subject in the kinematic study of brain activities. These topics are detailed below.

To begin, assume that the typical state of firing gate configuration in pathway A is excited to a topological state of B, with a definate frequency and demonstrated amplitude. During the enterance of a stressful input, which causes the layer to be strained beyond its linear range, configurations change from a typical topology of B into a typical C and finally in a typical D topology.

These changes occur after an interaction of incoming configurations with previous structural configurations and the resulting configuration tends to settle as D. These changes are illustrated below[1].

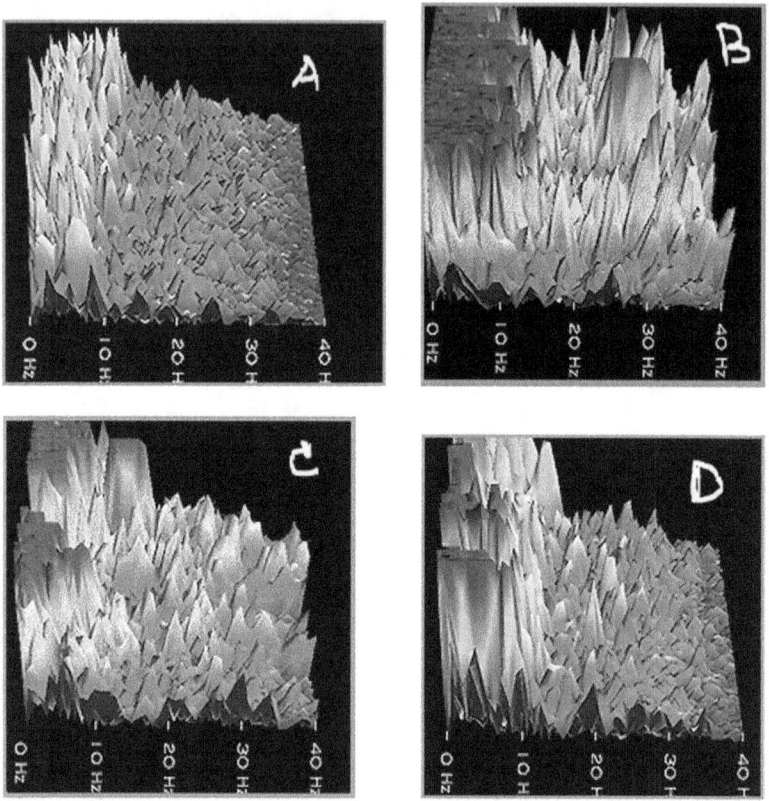

Configurations A and D are not identical; the difference is caused by temporary confined energy that is accumulated as strains in the given location. The confined energy is released in a regeneration period and results in a change in local configuration (permenant consolidated memory) and a current of energy release (emotional relief), in the shape of dreaming. The difference of the configuration energy between states B and D, or the

---

1 Adopted from: http://www.brainev.com/Research-Benefits/Comparison. aspx

accumulated energy, (shown in E^2) will be consumed in a motor action or a cognition process.

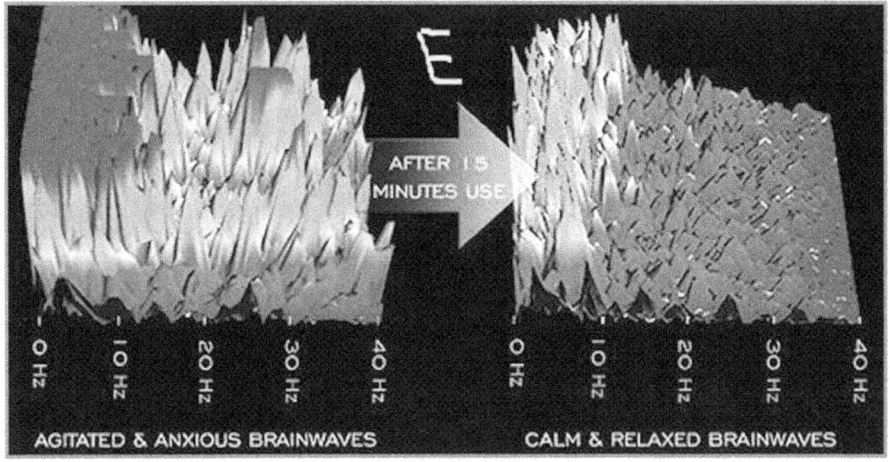

In short, shear flows along stimulated synchronized firing routes as the pathway; because of the shear confining-shear releasing (in elasticity terms: shear storage/ shear loss) imbalance the location goes under strain; by changes in input shears, strains would be released partially and absorbed for the remain; the remains of strains would be accumulated; the accumulated strains cause a decay in elasticity property of the layer; regeneration is a required process to consolidate or release accumulated strains to recover elasticity property of the layer.

Continuity of Structure, Function and Behavior in Different Levels of the Outer Brain Layer

There are different levels of energy transfer media in the brain. Pathways carry a developed energy wake from energy packets; substrates carry energy packets; neuron fibers carry lump eddy energy; and synapses carry firing strings.

---

2 Adopted from: "Brain Evolution System", http://www.brainev.com/ Research-Benefits/Comparison.aspx)

Individual firing intervals are influenced by molecular structure and chemical environments, while genes as components are one of the main sub-molecular catalytic elements which determine their functions. Synapse connections direct firing currents; fiber nets create energy orbits; substrates are bed streams for energy packets and layers host firing energy fractals which are growing together; and, when the growth is ended, a wake of energy is created. All of these energy flows are conducted in different states of laminar, transmittal, turbulent and chaotic conditions during a brain activity.

Synchronized firing currents are caused by increases in firing density in a substrate that is excited. Currents organized in rows produce tiny electromagnetic fields. Electromagnetic fields then synchronize the firings in the next substrates. Therefore, there is a two-way interaction in a pathway: synchronized firings progresses and electromagnetic forces is inducted and both influence each other.

On the synapse level, firings are influenced by inputs and attractors, both of which change the density, synchrony, lock-in-phase and configuration of firing layers. Therefore, the synapse level is under the influence of entering inputs, attractors, and energy currents in the shape of clouds of firings. The type of energy transfer is chaotic. By moving from the synapse level to the layer level, dependency on inputs reduces and energy streamlines are conducted by attractors. Therefore, the correlation between energy units change to a macro scale and the state of flowing changes from a turbulent to a laminar condition. Finally, on the layer level, the correlation between energy units is wake-like which is strong enough to stimulate a motor output signal.

Energy transfer on each level, from synapse to layer, proceeds according to the stress-strain properties of the carrying media. These properties are organized into three categories:

1) Structurally, a pathw2ay consists of several substrates; a substrate is made of neuron fiber networks and each fiber contains building synapses.

2) Functionally, there are many arrangements of firing clouds, which are connected in strings and configured in an orbit; then layers of orbits are configured in an energy packet and a mass of energy packets produces wakes of energies.

3) Behaviorally, these clouds possess a chaotic flow on the synapse level, a turbulent flow in lump eddy particles, a semi-regular flow in energy packets, a laminar flow in energy wakes and finally, a pulse-shaped flow in output signals, or commands.

The media or material of pathways, substrates, fibers, and synapses show different viscoelastic properties and present different characteristics of the stress-strain relationship. Therefore, in terms of saving or recalling memories along a pathway, different pathways have a functional time dependency that varies, from the very short to the long term. The higher the time dependency of a material or pathway, the shorter the term of the memory stored there, while material with a lower time-dependency will store long-term memories.

The growth of firing fractals during any brain activity proceeds from the synapse level and grows to cover a portion of the outer layer of the brain, limited to a pathway. Energy flows from sensory channel nerves to the input zone synapse level. Created firing fractals develop as energy packets in a substrate level and continue as energy wakes through the brain's layers.

Energy entering into a structure of synapses imposes stress forces on that structure, threatening to deform it. The structure is flexible, up to its elasticity limit; when the stress from the energy pushes the structure beyond that, the structure absorbs the excess of stress force as confined energy. This structural resistance against energy flows generates stresses, and the stress-strain characteristics in any level define the transfer function. Therefore, the direction of stress forces moves in opposition to the direction of energy flow. The interaction of these opposite forces, as operating and generating operators, alters the streams of firing gates' configuration to form a temporary balance.

The accumulation of changes occurring in the brain during the waking hours makes a given layer less elastic. Sleep is required to halt the entrance of new inputs and to recover the brain's elastic properties by releasing confined energies in the form of strains. Depending on the strength of an imposed strain, a memory of the strain-causing source would be saved as a short, long or permanent memory. The continuity of saved strains also varies from continuous to discontinuous types when going from the synapse level to the layer level. Accordingly, related memories are also categorized as a declarative or procedural memory, more static in layer level and more dynamic in the synapse level.

Kinematics Explanation of Sleep, Memories and Dreams

Sensory inputs during the waking time induce stresses along the pathways of processing. The inputs are entered and processed as a shear flow along their pathways by electromagnetic forces which are induced by synchronized firings. Low firing frequencies or low loading rates give a nonlinear viscoelastic property to brain tissues with lower storing and less relieving properties; while increasing in firing frequency and stress rate increases the linearity property of the brain's viscoelastic property.

Stiffness increases with increases in frequency, making an active pathway to be different from boundaries which have no synchronized structured firings. If it would be assumed that each pathway is made of substrates of segregated firing segments, a natural frequency can be defined for any substrate structured firings. The Brain, as a viscoelastic material, has elements of viscous and elastic properties and due to viscous property resist to shear flow and exhibit time dependent and phase lag to strains. Generally, a linear elastic behavior is expected for small sized substrates and nonlinear for compound substrates.

A summary of the strain-stress relationship for the brain tissues are as follows[3]:

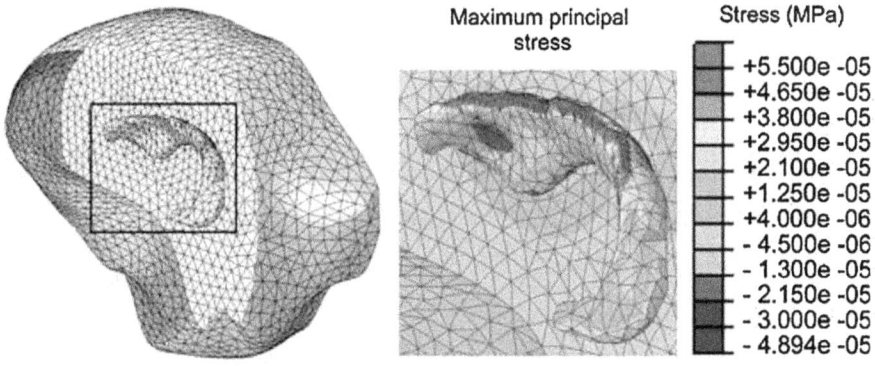

Fig. 3 Picture on the *left* shows a three-dimensional model of the brain and cerebral ventricles. The biomechanical stresses (in MPa) along the periventricular region in hydrocephalus is shown on the *right*. Positive stresses imply that the tissues are in tension while negative stresses imply that the tissues are in compression. Pictures are adopted with permission from Cheng and Bilston [12]

Although these properties are true for individual tissues, but the ranges are much limited for viva tissues in place and under normal brain activity loads.

- The stress changes from compression of -0.06 Newton/mm² to tension of +320 N/mm²

---

3 The figure and some physical data are extracted from the book: *Neutral tissue Biomechanics*, By Lynne E. Bilston, Springer: 2011

- The strain varies from %0.05 to %10; which it happens from %0.05 to %0.5 in the linear range and higher in the nonlinear range of the elasticity.

- Stress load frequency changes between 1 and 50 s^{-1} in a nonlinear range and lower than 1 in linear range

- Although the relaxation time for a fraction of tissues in a pathway due to connections to and influence by other tissues is not complete during waking time, because stress loads will not be removed practically up to sleep time. Therefore, accumulation of shears would be stored in locations, which change the stress-strain relationship accordingly. In other words, the above information can be approximated in a fraction of time and will be increased during the waking time.

- Strains are a function of stress load rate, type of the stress and accumulated load of stress as the confined energy

- Straining is linear for fine substrates, but nonlinear for integrated substrates

- Through the accumulation of shear loads in locations and loss of elasticity, the brain tends to not receive any new stress, which means it goes under the regeneration process.

The confined energies which had overstrained some location's tissues during the day will be regenerated by: 1- breaking some existing connectivity and making new ones (consolidation of a memory) and 2- distributing the excess of the confined energy in a sequence of substrates (dreaming).

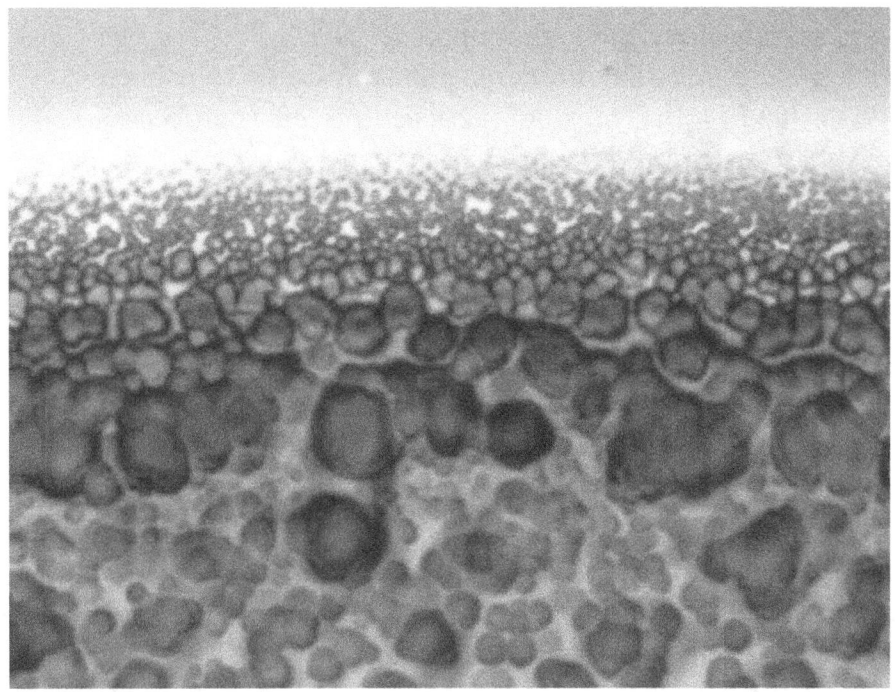

Landscape of a self-assembled morphology presenting typically how different sizes of energy packet relatively can form in a brain layer[4]

Pattern Formation in Brain Layers

Soft materials such as polymers can self-assemble into regular patterns by merely changing ... [one of main influencing parameters] of the system. For tri-layer films in which the central layer is much softer than the two capping layers, we have observed the self-assembly of a novel, periodic, in-plane morphology for tri-layer films consisting of a wide variety of materials combinations. We have developed a simple theory, based on linear stability analysis, which correctly

---

4 From: *Pattern Formation in Polymer Multilayer Films*, Department of Physics, University of Guelph

predicts the dependence of the wavelength of the mor-phology on the individual film thicknesses, as deter-mined by the interplay between the energy decrease associated with the attractive dispersion force acting across the tri-layer film and the energy increase as-sociated with the bending of the solid capping layers.

The morphology ... [will] be removed by altering the dispersion interaction, and can be reformed with a larger periodicity.[5]

The above statements are the results of intensive laboratory research conducted in Department of Physics at the University of Guelph. The brain is a similar material as that described above, thus it exhibits similar behavior by forming patterns in layers and having strings of synapses in nerve fibers which are bound in nodes. In addition, neuron fibers carry an electrical current in case the firings that occur along their synapses are synchro-nized. A synchronization event is a self-organizing phenomenon in which incoming stress, initial conditions and the influencing parameters are looped in a cycle linked by delayed electromag-netic field creations. The inducing stress introduces a wave of force fields that periodically strain and release the media along the pathway.

If the inducing stress is a stepwise constant input, strain will be time-dependent, as illustrated below[6].

---

5 Ibid

6 From: *Viscoelastic Properties of Vimentin Compared with Other Filamentous Biopolymer Networks*, by Paul A. Janmey, Ursula Euteneuer, Peter Traub and Manfred Schliwa

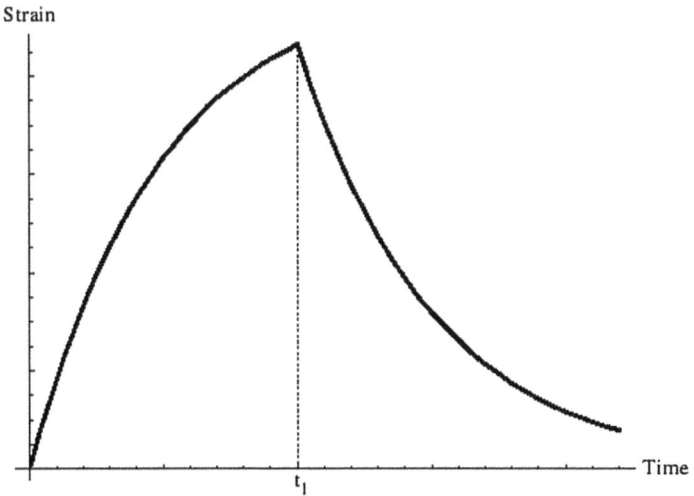

Typical strain changes with time for layer of fibers

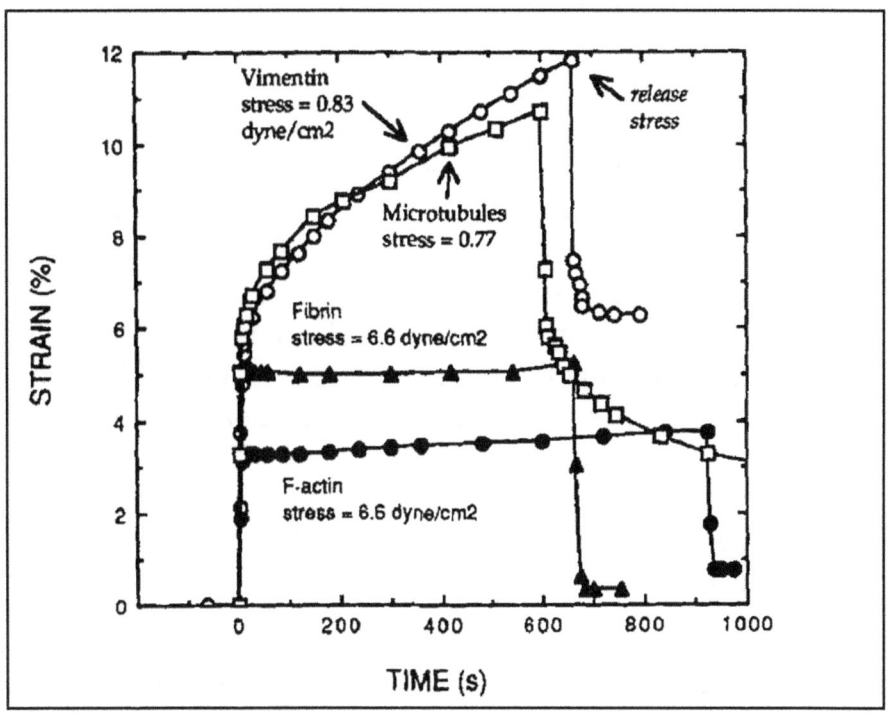

Behavior of different polymer networks under a constant
stress rate of 1 dyn.cm^{-2} s^{-1}

If the inducing stress occurs as pulse inputs, the pathway undergoes an initial strain, which gradually decreases to a residual strain.

The selection of the substrates under strain depends on the required activation energy, which each substrate requires to let the incoming energy in. For a selected pathway, if stress is below a critical value, the pathway shows a linear behavior of viscoelasticity. A plot of the ratio of stress per strain relative to time indicates an independency to stress at this range and, regardless of the stress initiator, a similar family of patterns will be created in reaction to this level of stress. However, in the non-linear range, configurations will temporarily be slightly deflected, thus retaining a trace of the stress source. When a stress load approaches that limit, there is an instant of deformation as a consequence of elasticity term (σ) hosts an energy packet. The substrate shows inertia toward any entering energy, which tends to change the substrate's configuration.

The angular velocity of a packet of energy is the rate of its rotation when the packet enters a substrate. The angular momentum of a pathway is the summation of the products of two different vectors: first, a position's vector and second, linear momentum vector. The related forces are induced by electromagnetic fields. A developing pathway in the brain is a circularly polarized plane formed by the resulting forces of tiny electromagnetic fields and carries an angular momentum. The angular momentum of a given volume of the free electromagnetic fields has three components: spin integration, orbit integration and layer integration,[7] each representing, respectively, changes in the fibers' angular configuration, eddy and packet energy contents.

---

7 Stewart, A.M., "Angular momentum of the electromagnetic field." The Australian National University, 2005. Arxiv.org/ftp/physics/papers/0504/0504082.pdf

The reason that a pathway becomes a circularly polarized plane when excited is that, in addition to the energy flux direction, energy units in different levels have different directions of random motion and so induce angular momentums on each other. The kinetic energy of the massive energy eddy particles or energy packets, which rotate in position, is proportional to the square of the variable angular velocity.

It is hard to say that sensory inputs are entering with external torque. Therefore, the brain can be considered a closed system and therefore angular momentum conservation applies to it. The conservation law can help to explain many diverse phenomena in the brain, like the formation of the brain's folds.

To apply the energy conservation rule, when kinetic energies are confined and consolidated as potential energy, which is then partly crystallized into new configurations, the diverse phenomena referred to will develop.

Energy Level and Concentration Degree

Energy level, at any point, is a function of point frequency and amplitude, and it can possess the variables of frequency and amplitude in different combinations. The general equation that is most often used and provides a reasonable measurement of magnitude is:

Energy factor = $(frequency)^a_*(amplitude)^b_* 1.6*10^{-16}$, when the conversion factor of 10^{-16} is a conversion unit from femto-electro-volt (fev) to joule.

The other parameter that can be used to indicate different dispersions of energy transfer as it pass through different layers or areas is called the concentration factor. This expresses the degree to which the deviations are arranged in flows that align with or against the dominating flow or the degree to which the different flows are integrated into an outgoing complex.

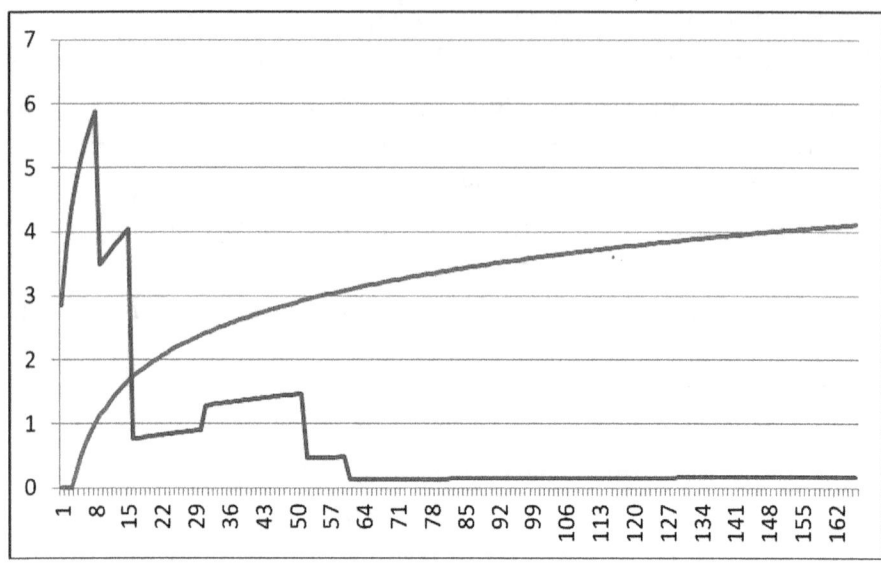

Energy factor and concentration degree versus frequency (concentration factor would be the highest for highest frequencies which occurs in the frontal lobe and lowest in activities happen in the inner brain)

The concentration factor has less value in lower frequencies, which represent energy flow that travel through inner is less spread, but increasing in middle or outer layers, and finally the highest in the frontal lobe. In contrast, the potential energy level is highest in the inner brain and reduced in higher layers, progressively; however, it is higher in the input-output zone, as compared to other areas of the outer layer. The kinetic energy level would follow the reverse order of potential energy levels. The frequencies and amplitudes used to describe different states and different layers are taken from data collected by laboratories measuring EEGs, or other methods, and are categorized according to the following ranges of frequencies:

Gamma in a frequency range of 60-135 Hz: this represents frontal lobe activities, and is involved in a certain cognitive or motor function in super-conscious states.

Gamma in a frequency range of 30–60 Hz: this represents the combination of energy flows in the frontal lobe, and ending in a certain cognitive or motor function in high analytical-synthetic states.

Beta in a frequency range of 12 Hz to about 30 Hz: this is seen toward the front of the head and in fully conscious states.

Alpha in a frequency range from 8 Hz to 12 Hz: this is seen in the posterior regions of the head corresponding to middle brain activities

Delta in a frequency range up to 4 Hz: this displays the highest amplitudes and the slowest waves; it is seen normally in sleeping adults, especially in NREM stages.

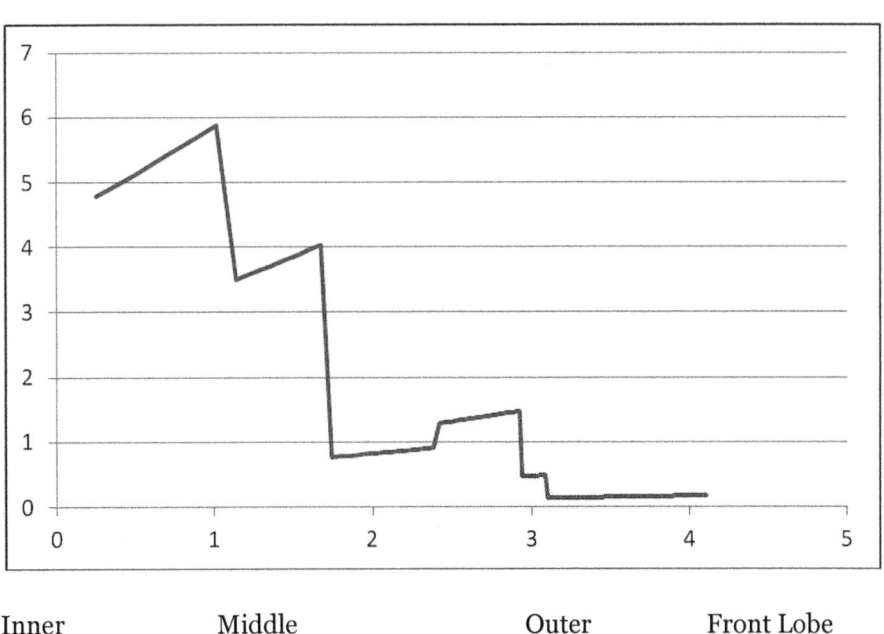

| Inner | Middle | Outer | Front Lobe |

Concentration Degree

Outer Layer Regularized Oscillation and Output Energy Signal

The following model provides a brief description of how flows of energy work as a media for any brain activity. Firing fractals grow and couple with other firing orbits and fractals, developing into a larger synapses configuration. Any local firing fractal that is added to the incoming fractals increases the regularity of energy layers' shapes, and their oscillation rates. The stimulation of the initial substrate causes the energy orbital to be created and to grow in a firing fractal until it reaches the limit permitted by the similarity in structure.

Families of firing fractals that have a general similar shape are created in similarly structured regions. Firing fractal growth is limited by overall synapses' configurations. The well developed energy fractal conveys an outgoing signal to the motor outputs (spinal cord route). A firing fractal is disturbed and finally disappears when the coupling between the fractal orbits disturbs or when a dominating input is no longer available. By contrast, regularization of the layers of energy in a firing fractal is increased by the increase if attractors' activation. The terminating point for a grown energy fractal is in the output zones.

Attractors as the seeds of growing firing fractals, emanate from a variety of human needs, including those associated with the physiological, social and safety, as well as from the need for self-esteem and self-advancement in respect to existing or imagined circumstance capability to satisfy a need. Attractors are located anywhere appearing as a circumstance imagination by static characters and their moves reflecting an emotion which shows how much the circumstance satisfies a need. The higher the number of attractors, the more complex the associated brain activity will be.

In addition to inputs and attractors to initiate firing fractals, firing fractals can develop from any overstrained point on a layer, when the sensory input is absent. In any layer of the brain (inner, middle layer regions, frontal lobe), a firing fractal can be initiated, develop and make changes in synapse and fiber configuration either temporary during the waking time or permanent during regeneration period of sleep.

Output motor signals are either an intermediate or final product of a developed firing fractal and these signals are replaced with a new growing one when their energy is discharged. Intermediate products communicate with new inputs as feedbacks to create a more comprehensive output. Such iterations continue until an intermediate product would be boosted to the required activation energy level to send out a signal to output channels as a commanding action.

This model closely resembles the model presented by Larry Abbott at the "From Quantum to Cosmos" conference, held at Waterloo University in 2009. His lecture on the model is titled "Sense from Chaos: Controlling the Dynamic Networks of the Brain." Like the model being laid out here, his model describes the way in which a state of regularity in firings can be created from a state of chaos via slight changes in firing states, like changes in spike rates and frequency occurring in a segment of the brain during a brain activity. Such regularity in firings appears in the shape of a fractal.

Discrimination Between Sensual and Internal Inputs

Inputs entering and initiating brain activity are limited by: 1) input channel structure, 2) pathways' stress absorbing and transferring limitations, which is determined by the layer stress-strain characteristics, and 3) the output of the created signal. Therefore, energy flow limitation is determined by the boundary-limiting inputs and outputs, and that boundary is identified by the neuron fibers' overstrain. The stream of information energy entering is too weak to be measured; however, energy flow is boosted by nutrition energy. There is a continuous competition, between entry inputs and stimulating internal attractors, to initiate brain activity; whichever of these competitors that first surpasses the required activation energy needed becomes the initiator of brain activity. Attractors are consolidated structures of synapses and fibers found in regions other than the outer layer; when excited, they direct the pathways in the outer layer. Attractors excite the expectation of their targets. A feedback of deviations to the pathway direction enters as an additional input to correct the direction of the energy transfer in the new run. The flow of free energy continues until a new balance is established.

In addition to the source of initial stimulation, the singularity or non-singularity of inputs initiating a unique brain activity must be accounted for. A working memory of a sensory input has a life time of 200-500 milliseconds after perception; sensation channels should have a limitation that permits different bands of frequency to enter during the same period of the time. In statistics, the reliability of a random variable measurement among several other variables is indicated by the Fano factor, otherwise known as the noise-to-signal ratio. The Fano factor indicates whether or not the singularity of inputs for a distinguished brain activity is based on a correct assumption.

In short, in the rush of entry inputs that occurs as environmental inputs are trying to input, included will be some the specific inputs requested by the active attractors. However, only those that can overcome the required activation level have a chance to enter. This means that the general state of all ongoing activities make the brain highly sensitive to selected inputs. However, brain function in this regard is not determinative or predictable; therefore, a cause and effect formulation is mostly not applicable to brain behavior.

External inputs can impose an order on brain activity, or they can disturb an existing order. There are different types of inputs produced by the environment, such as jet-type, step-type and random or trigonometric functions for forced inputs. Some of the types help to produce a longer continuous and more regular energy flow. Others induce irrational brain activities that are not connected to the initial energy flow. Jet-like inputs induce cycles of storming and initiate an activity far different than alternative-stepwise inputs, which occur either as instant shocks, short or long-duration stepwise inputs, or as trigonometric functions according to the frequency of occurrence.

From Circumstantial to Subjective Insight

When brain activity shifts the state from feeling to automatic thought and then to thinking efforts, mentation content changes from qualitative (circumstantial) reaction to semi-quantified (subjective) cognition and finally to quantitative (very subjective) action. The degree of focus will increase in the same direction, moving from a minimum to a maximum possible level of attention. In a feeling state, the pathway develops in a unique way and due to its lack of branching, overstrain is highly likely to occur in the location of such a pathway. In automatic thoughts, diverging pathways with sequential focus develop. In a thinking state, converging pathways emerge and are synthesized in the frontal lobe.

There are two different ranges of stress impact on pathways; these are called linear and non-linear ranges of stress-strain correlation. Intentional focus occurs in the linear range, while a brain activity in the non-linear range is accompanied by subconscious focus and emotional loads. Consequently, it can be said that stress load is a function of the degree of emotion and focus. Physically, any area in the brain displays a different type of elasticity character, depending on the type of neurons that build the area. Middle layers may have more lattice structures to store these memories and be suitable for names, figures and event facts, while outer layers are structurally linear and typically suitable for functional behaviors. Thus, highly emotional loads, which have large energy fractal amplitudes, are routed to the middle of the brain, while focused correlations can happen frequently in the outer layer, especially the frontal lobe. Normally, lager amplitude energy packets are less dynamic and, therefore, possess a lower frequency.

Frequency, as a parameter, is closely related to any structure strength. Therefore, the characteristics of brain activities can be traced using the frequency of the pathway structure as the bed stream for its activity. Energy packets with lower frequencies and higher amplitudes are absorbed in the middle layer, while the energy packets with medium frequency migrate to the outer layer and those with high frequencies and fine amplitudes move most often to the frontal lobe.

The stress impact, which depends on amplitudes, can be plotted against the focus property of activity in the area. Assuming zero focus intensity for the unconscious state, which is related to activities in the inner brain, the focus intensity of activities, which proceed in other parts of the brain, would be related to the reverse of frequency *(logarithm of (frequency) -1)*. Therefore, a super-conscious state has a focus intensity of approximately three; for a conscious state, it is approximately two; for a subconscious state, it remains less than two; and during an unconscious state, it is approximately zero.

During a regeneration period, the release load from strained locations is equal to the total confined energies, reduced by the energy consumed in the synapses' reconfiguration. The released confined energies excite the substrates on different levels of the layer around the strained location, recalling their associated memories. These memories are perceived as symbols, metaphors and similes and are wired into a narrative in a dream state. The firing fractals created at the synapse level create static images like shapes, names or figures; where, energy patterns at substrate levels create dynamic images like motion, speed and acceleration. Both levels' are governed by the function specialty which is imposed over them through the overall layer. Any input starts with a static concept (a player), such as a person, a substance or an animal. It continues with the player's behaviors, actions or reactions, and then creates the scenario of a dream by connecting static concepts to dynamic concepts. The input is initiated by the release of strained points, while the same pattern that was being strained during the waking day is reproduced. It is an iteration process that grows as a fractal. The iterated core is the same strain pattern that reflects the theme of a given emotion. In other words, the theme of an emotion remains the same during the strain course, and it is the same for the recalled memory and emotion that occurs during dreaming. The sign of the emotion theme can be positive or negative and it can correspondingly work as an excitory or as an inhibitory component in the flow of dreaming. An excitory (positive) emotion theme will produce delights in the dream, while an inhibitory (negative) emotion theme results in a negative state of mind.

Tracing the pathway of a narrative and memories recalled can theoretically be achieved by measuring the chemical Norepinephrine's level in different areas and comparing the measurements with the Norepinephrine level in the Reticular Activating System, which is found in the brain stem (or the attention center). The concentration of this chemical in any area indicates how active a given area is in a dream.

CHAPTER II

Growing a Firing Fractal into an Energy Current in the Course of Brain Activity

The parameters of an active pathway are determined by its inputs, boundary limits, functional properties, and the attractors which influence its direction. If the initiator of the activity is considered as an input (either a sensory input or an over-strained location) and the elastic property defines the functional properties, the surrounding substrates, which do not approach the activation energy required to initiate brain activity, build the boundaries of the pathway. Considering a fractal growing in the course of a brain activity, and to imagine the reproduction of units to grow the fractal, split the frequency and amplitude of the input energy by functional property scaling down for a factor of "r as the fractional ratio." This split produces "n" number of self-generated inputs. The produced particles are then integrated in one small packet. The scaling down and integration process is repeated for the larger packets and eventually develops first into a pathway and then into a layer. The pattern may be repeated in infinite dimensions on the scale of synapses' firing gates, but a developed pattern on the layer scale takes on a definite shape with definite boundaries and possesses a finite and enclosed amount of energy.

Pathway limits are defined by the substrates activated when the entering input produces the required amount of activation energy. The amplitudes of substrate configurations are random. The boundary limits of the substrates at different levels of synapses, fibers, eddies, packets and layers can be considered to have a scale down ratio of "d." Therefore, the boundary limit of a layer level defines the smaller boundary limits in lower levels repeatedly. The amplitude found in different levels is random; it is distributed and scaled by a factor of "d," which is related to the original "d" and is formulated this way: $d_1 = (\frac{1}{2})^{H/2} d$. The center of any area within a boundary limit, in any level, can be determined by finding the average of all the boundary points of the activated configurations and adding that average to the random variable of the energy packet defining the higher-level amplitude. This process can be repeated for several levels, from the synapse level to that of the layer, or from the micro scale to macro scale, and for different sizes of energy particles and packets.

The value of H defines the smoothness of free energy transfers through layer levels. As H increases, smaller integrating energy particles lose their individuality and the overall pattern becomes more integrated into a whole. By this description, a landscape of an activity in the brain (a mentation) can be pictured as starting from individual, fine firing fractals, then evolving into energy packets and finally creating energy wakes in a layer.

Mentation Stability and Duration

In physics, the media for particle interactions are called fundamental particles. These carry forces, fields and interactions. Beyond a threshold limit of forces, atom components and, later, the structuring of atoms, molecules and more complicated structures develops. The loss of force between integrating elements, beyond the limit necessary to maintain rigidity, again breaks down and the structure collapses into sub-units. In a stable structure, interacting forces' strength oscillate regularly, while in an unstable structure, interacting forces oscillate irregularly and weaken the bond; finally, the interaction between elements collapses when oscillation becomes too irregular. In addition, each particle has an angular motion, which is indicated by spin. Particles with different levels of spin, which measures the differing polarity of their frequency-amplitude distributions, produce a strong combination and results in a normalized distribution. Therefore, particles with different spins attract each other and produce a coupling. Two exposed particles with the same spin increase the polarity of their frequency-amplitude distributions when combined, and therefore do not attract each other. The interaction mechanism is normally described based on the creation and disappearance of regularity in the frequency-amplitude distribution of fundamental particles during coupling.

A similar behavior can be observed in energy packets using the patterns they create through their combinations. If energy packets' synchronized firings frequencies find a correct spin, they grow as a fractal. If their synchronized firings frequencies do not follow such a regulation, they do not develop a regular shape. In short: given a normal frequency distribution, the maximization of amplitude dispersion is equal to the disappearance of the known firing configuration, and the minimization of amplitude dispersion result in the creation of a fractal pattern of firings.

Each energy packet is a complex formation of eddy fractals, all fluctuating at a variable frequency with the same amplitude;

this fluctuation exists between a given maximum and minimum range of frequencies. Meanwhile, the energy packet's fractal, itself, has an increasing and decreasing trend of amplitude dispersion found in an energy wake in the overall layer. In other words, any pathway of brain activity is regulated by layer functions as well as by energy packets and substrate functions. Similarly, each energy packet follows the rules of a pathway as well as a firing fractals rule. The general rule governing both of these processes is summarized by the stress-strain correlation of all three levels: synapses, substrates and layers.

On the micro scale level, electromagnetic field and subsequence force field development is the result of the regularity and synchrony of firings and firing gate fluctuations. However, couplings and interactions are combinations of energy units that occur on different scales of firings, eddy particles and energy packets. Depending on how many couplings happen simultaneously and how many couplings are strong, their stability is determined on a macro scale; that is, the higher the number of couplings, the higher the stability. The way in which couplings occur on the micro scale determines the functional or procedural properties of the location of the related internal interactions and external behavior.

The micro and macro interactions between energy particles appear as form-like, or wave-like behaviors of the matter. Relations between form-like and wave-like appearing can be described in "time and energy" terms. Because energy is related to amplitude and frequency is the number of happenings per unit of time, or the reverse of time, any energy particle alternates pulses between inertia and moving states; hence, they can be analyzed on the basis of oscillations by measuring for frequency and amplitude. Meanwhile, the spin of the energy particles involved in an interaction indicates the stability of a given coupling, as described above. Consequently, any brain activity can be understood based on the combination of the building energy packets' frequency and their irregularities as measured in the whole of the activity's energy flow.

A brain activity's continuity is studied based on the creation, duration and disappearance of its building components in different levels of firing, on energy fractals, as well as on the energy wake in the layer level and energy packets' continuous spatial and sequence frequency maximization and minimization. Through their spatial direction, the combinations of amplitudes (as a measure for pathway length) and frequencies (as the natural frequency on the combination) describe the creation, stability and disappearance of a brain activity. These parameters depend strongly on adsorption and desorption; or, in other words, they depend on the media elasticity of the pathway through which energy particles travel. The difference between energy absorbed and desorbed accumulates as a confined energy in waking time, and this degrades the elasticity of a pathway. The confined energy must be released to recover the elasticity property during sleep. This phenomenon is the basis of memory saving and memory recall of brain activities and their initiators.

Longitudinally through a pathway, a typical *amplitude-frequency distribution* goes through structural, regular and semi-regular transforming, as well as chaotic form initiation before the disappearance of a mentation.

Transference between structural to chaotic states can be measured by the kinematic-to-potential energy ratio (the distance from equivalency) for the consisting energy particles. This ratio should change from a low value to a high value, indicating the formation of the fractal; at the highest value, the fractal finds a very high stability degree and then decline to a very low stability degree by reducing to the lowest value in activity termination. A functional operator creates a prefect structure, but gradually the structure would melt into a behavior and disappears. In terms of cognition, an energy configuration, as a static image, converts into reasoning or a feeling or an understanding and then the configuration disappears. The distribution of the frequencies, against the amplitude of the fractal components, gives an idea of the stability and duration of the present brain activity.

Config Type (y-axis) — Distance from equivalency (x-axis)

Columns read vertically: processing / structure / regular / motion / semiregula / motion / turbulency / field / disapearin

A Summary of the Fractal Study of a Mentation

A mentation is attached to a firing fractal growth, made of firing orbits, which convey a perception, a feeling, a memory recall, a thought, or a dream. A mentation conforms to all the properties which identify a pattern as a fractal,[8] because the media that carries it is as follows:

1. It consists of synchronized firings that are copied in different levels of a substrate's structure, which can be retrieved or saved.

2. It consists of synchronized firings are too irregular to be described by Euclidean geometry.

3. It consists of synchronized firings that are based on very fine structures of ion channels and synapses, as well as macro structures of substrates and pathways.

---

8 Falconer, Kenneth (2003). *Fractal Geometry: Mathematical Foundations and Applications*. John Wiley & Sons, Ltd.]

4. If the oscillating amplitudes of the components are considered as their particle size, mentation follows the dimension rules governing fractals.

5. Mentation can be well defined as a free energy transfer.

Ion channels, synapses bundles, substrates and pathways build the position oscillators; and firings, strings of firings, force wave packages, momentum energy pockets and energy wakes make up the energy particle oscillators. The interactions between mentation components are carried out by positions and the energy particles that pass those positions.

Generally, A pattern of synchronized firings at different scales (which are: 1- outer brain synapses bundles, substrate and pathway levels; and 2- inner, middle and outer layers) creates bifurcation fractals that have self-similarity and are iterated by inputs so that a quasi-equilibrium is found between them. A memory or need (physiological, social, safety, or self-esteem and self-advancement) influences a pathway in the outer layer by growing as a bifurcation fractal and resonating by processing input.

Therefore, wake energy patterns in a pathway, energy packets in a substrate and eddy patterns in a fiber segment (between nodes) are copies of each other; their natural frequency is a factor of each other; and they are synchronized together, either in saving or recalling communications. Communication for local-continued routes is achieved through momentum transfer and, for far locations, resonance facilitate communication.

A wake energy pattern can be imagined as a complex fractal made by the reproduction of firing fractals in substrates. This type of fractal formation is known as bifurcation, because at least two different frequencies are involved in different areas.

The energy fractals appear in any level or layer imposing stresses over the structural configurations of synapses, fibers and their network and they tend to enfold the location as a whole.

The temporary deformation resulting from energy transfer follows the stress-strain properties of that location. However, the temporary deformation will be recovered for overstrains in regeneration periods.

Brain integration dynamics

Brain functions can be analyzed based on the following parallel dynamics:

1) Strain-stress time lags behavior in different levels of: 1.1) neuron fibers; 1.2) homogenous substrates and 1.3) brain layers, depending on the input strength and type. Strain-stress correlations in absorbing and desorbing periods are time-dependent. Desorption in the nonlinear range occurs well after absorption is complete and the time difference depends on the elastic behavior of the media.

The brain, as a viscoelastic material, shows some phase lag in the resulting strains, which are caused by a stress overload. If the stress load is an oscillating type $\varepsilon = \varepsilon_0 \sin(t\omega)$, the resulting strain will be $\sigma = \sigma_0 \sin(t\omega + \delta)$. Where, ω is the frequency of strain oscillation and δ is the phase lag between stress and strain. Any strain consists of two components: elastic and viscous properties.

The two relevant transferring energy portions are: 1) saving and releasing $E' = \frac{\sigma_0}{\varepsilon_0} \cos \delta$ and 2) a change in structure (e.g. sinusoidal slippage of fibers $E'' = \frac{\sigma_0}{\varepsilon_0} \sin \delta$ in their wrappings) meanwhile, the viscous portion is created by the configuration energy transmission from one substrate to next. Accordingly, the saving and releasing of the confined shear stress, as related to confined and released energy are: G' and G'' in any instant. The total energy and stress forces during an oscillating phase is: $E = E' + iE''$ and $G = G' + iG''$ in sequence. This means that there is a phase time difference between elastic and viscous-related stress force and energy.

In summary, the deformation caused by input forces, either external inputs or internal initiators, is composed of two parts: elastic and viscous elements. If the input force is constant, deformation due to elasticity remains constant; but deformation caused by energy transfer will increase until that inducing force ceases. When the inducing force is removed, the deformed elastic portion returns back to its original position, but the viscous portion does not. Therefore, the viscous component of the stress remains in the substrate. Due to the self-organizing nature of the brain, an inducing force, as a combination of external stimulation and internal resonances, remains in balance so that the self-organizing property of the brain is preserved. The distribution of shear stress, shear strain, and shear time-rate differs substrate by substrate, area by area, region by region and, intensively, layer by layer. The relation between the time-varying stress forces and energy transfer velocity reflects both the elastic and viscous properties of any location.

2) The energy, which transfers between three levels of fibers, substrates and levels, has different topological configurations, due to the influence of different types of shear forces in each level. Consequently, the energy flows in each level seem to be distinguished from each other and to possess a time-lag relevant to others, although all are travelling in parallel. The time lag result from the different sizes of energy eddies in those currents.

3) Electromagnetic fields decay as a result of inducing forces on fibers; therefore, there is a time lag between the proceeding of synchronization, electromagnetic field creation and the consequence of momentum transfers.

4) Aside from sensory inputs, the other source of brain activity initiation is internal resonances. These are self-inducing force fields that are self-organized and display the

properties required for the creation of energy pattern fractals. In the brain's pathways, the configuration energy is the dominant flow, but in output zones energy is decomposed into images of the environment and events, as they would be retrieved from related memories in the middle brain.

Similarly, when inputs enter the input zone through sensory channels, patterns equivalent to environment substances, which are sensed through relevant sensory channels, will be mapped by synchronized firings in input zone. Synchronized firings in a substrate integrate an energy packet with definite configuration which is called the configuration energy. The retrieving of the resulting images comes after the energy transfer is terminated and a steady energy wake is formed in the layer. The image attached to a final energy wake has a time lag in relation to energy transfer occurrence, due to the retrieving process in the output zone.

Considering all these time lags, stimulating inputs and the resulting patterns produce a total variable time lag, which causes the parallel stimulating of different inputs to be possible, thus initiating a more complex brain activity. In humans with developed frontal lobes, all these parallel energy fractals integrate, making a complex that represents conscious efforts, while in previous species of humans, such a number of parallel fractal images was not possible.

5) The configuration of an energy fractal in the brain is a vector combination of oscillating and transferring energy directions with spin on different levels of: firing gates level, the synapse level, fibers level, substrates level and layers level. Therefore, any eddy, packet or layer of energy has a property component of spin, in addition to its properties of inertia and velocity. In any level, these energy particles can band together if they have opposite spins; otherwise they expel each other. In other words, they show excitory and inhibitory effects according to their spinal signs.

6) Daily stored strains are accumulated and temporarily cause the deformation of synapses as well as flowing to make the energy through the layer uniform. The temporary deformations in nonlinear range would be consolidated and released during sleep to recover the normal viscoelastic properties of the brain as existed the day before.

7) Therefore, the integration of timely and daily dynamic behaviors of brain components, in different levels, strongly influences future brain processing and thus creates short and longer-term behaviors of the individual as a whole. If the fractal patterns are pictured in different fractions of time, one would find a general similarity between the patterns created in different levels, which is the characteristic of a chaotic media like brain.

With any initiation, the patterns change in a very chaotic manner, but with time they develop regularity, creating equivalent conditions between input energy and the appearing fractals, as outputs. From the initiation to the termination of an activity, fractals develop a broad family of patterns and are finally refined into a specific product. This process reveals the way in which kinematic energy changes, finally, into to potential energy; kinematic energy represents all transfers and potential energy represent all stagnant, confined and ground energies.

Family patterns include groups of patterns that have the same overall configuration, with only slight differences. The images attached to family patterns are symbols of each other, or they can combine in a metaphoric statement conveying the same feeling. These are the most internal processing products, but when they develop specific patterns they retrieve specific images.

Consciousness and super-consciousness is assumed to be a consequence of possessing parallel energy fractals, and a balance between their integrated complexes and individual energy fractals. Super-consciousness is a phenomenon that happens in the frontal lobe when the number of integrating energy fractals moves beyond a certain limit.

Two energy particles, at any level and with opposite spins, will band together to create a more complex configuration. If substances are reviewed from those with simple to those with very complicated shapes, it is obvious that the number of subunits and bands in subunits are increased as the substance become more complicated. Similarly, if we consider animals, energy fractal complexity reveals tremendous difference when compared between species in the evolution tree. In more complex animals, the upward trend of increases in new bonds, new subunits and new organs is exponential.

The brain, as a centralized integration center, has different areas and regions, and each has its own integration functionality; the more advanced areas take on the more complicated processing. However, the brain, as a self-organized system, is able to receive, process and resend information as well as to upgrade its own structure. Therefore, it is capable of structural development.

Application of Bifurcation Analysis in the Brain

The relation between a substrate and its related energy packet can be analyzed by finding its bifurcation point. Bifurcation analysis studies topological changes in a field, such as in the momentum force field in the outer layer of the brain. Momentum forces in a viscoelastic media cause oscillation, and frequency is used to establish the measuring parameter for these forces. A small change in the frequency parameter (like a disturbance in oscillation synchrony) or frequency number causes sudden change in the quality and topological behaviors of a pathway. Since the brain is a discrete system, these changes can be demonstrated by mapping the changing patterns.

Bifurcation can be analyzed by studying the stability properties of local periodic orbits of energy packets after their frequency goes beyond a critical limit. In addition to local equilibrium stability between substrate configurations and energy packets, differences between the imposed frequency and the natural frequency of locations can cause a bifurcation phenomenon to happen.

There are bifurcation points that increase order and bifurcations that lead to chaos. In the brain, a frequency change normally causes a change in the stability of a configuration state and consequently causes a local bifurcation. Following that, some topological changes in the neighborhood may regain the previous local equilibrium condition again. Stable saddle-nodes and folds in the brain are the result of the bifurcation phenomenon. Bifurcation points determine a substrate's structure resistance to energy packet configurations travelling through it and bifurcation occurs in ongoing energy fractals through different transfer levels and locations.

Application of Chaos Theory to Mentation

A mathematical expression may be able to define an energy fractal; if it cannot now, it will probably be able to in the future. The fractals involved in brain energy are limited to synapse configuration, and are therefore limited in number of verity.

I) Patterns based on synapse configuration in the rest position:

1. Initiator (in input zone locations)

2. Firing gate distribution describes the media-making synchronized firings fractals spread across a pathway by activation.

3. The process of synchronization in firings, developing from one substrate to the next along a pathway from initiation to termination, depends on the sensitivity of substrates to activations.

4. The scale of fractal growth depends on the repetition of self-similarity in the structure of neurons in a location. The number of the fractal dimensions in a location can be determined by the actual self-similarity limit in local structures.

    4.1 With the development of synchronization on different levels—of synapses, fibers, fiber bundles, fiber bundle networks and layers—a self-similar pattern of orbits or fractals will be created.

    4.2 The time span in which the fractal formation keeps its form depends on the duration of potential inputs and supporting attractors. However, the energy distribution enclosed in any substrate during a fractal's appearance is finite and cannot go beyond the substrate's elasticity limit.

5. The fractal energy content is limited to the self-similar growth of fractal forming units.

6. A fractal's clarity depends on a random value, corresponding to the viscoelasticity property distribution of the pathway and to the scaling factor, which measures to what extent self-similarity is perfect in the pathway.

II) Disturbance of energy fractals by changes in initiator

1. Firing gate distribution serves as the streambed for the current of synchronized firings. Because of changes in current directions, tiny electromagnetic fields will be created through the energy travelling route. Accordingly, the em fields impose forces on the streambeds of synapses, fibers and the pathway. Imposed forces result in slight changes in the firing gates' arrangements. The rest positions of firing gates resist forces that would change the position beyond

the location elasticity limit; but, regarding the changes that occur within the limit, the configuration energy content of the substrates will be altered accordingly.

2. Changes in local configurations can be considered as the travelling energy.

3. The interaction between ground energy (firing gate position distribution at rest) and travelling energy (firing gate position distribution after the imposition of em-field forces) encompasses the materialistic process of the fractal, from appearing to disappearing.

4. The dynamics of interactions are accelerated by the resonating influence of attractors, which have a natural frequency of the same or multiple frequencies of coupling substrates. The recalling and saving of memories is expected to be carried out by the resonating effect and the growth of an energy fractal. An activated-saved pattern in a far substrate is then combined with the main stream of transferring energy in outer layers, proceeding according to the rules of scaling, distribution change and axes rotation. This process will be repeated for all attractors, which can resonate with the main fractal.

5. At the conclusion of this process, the total interaction of fractals in a pathway creates a new complex fractal, or decomposes to some simple shapes.

6) All of the developing synchronization fronts are attracted by self-similarity rules and active attractors.

III) Saving a memory of the resulting energy fractal
The concluded energy fractal will convey the energy as a commanding signal to an output zone for expression. The remaining energy is trapped as strain or confined energy, and is

balanced by synapse reconfiguration and distribution in order to make uniform the energy throughout the layer during sleep.

Geometrical Patterns as an Edge of Chaos and Fractals

The brain receives outside messages by way of geometrical patterns that are translated by synchronized firing configurations, which initiate a new energy distribution in a pathway. The transferred energy settles down into a balance situation as a geometrical pattern that shows a definite configuration of synchronized firings through the pathway when stabilized into a wake. The produced wake is then absorbed as a memory of the formed wake and will be exposed back to the environment as an expression or commanding action for the other part. However, the energy pattern can be discontinuous in the scale of firings, but it is continuous in the scale of space that it covers[9].

Emotions as Attractors

Although conscious mentation is distinguished from feeling by its fractal complexity, it has emotions at its core. A conscious mentation is made of complex feelings. Emotions are created in the course of mentation and are the attractors for mentation continuation. If a mentation is positive in intent, it is excitory, and if negative, it is inhibitory.

In addition to feelings, all conscious and super-conscious processes—like reasoning, problem-solving and settling on a solution—are attached to emotions. Emotions are the drive for any kind of mentation in the brain. An emotion's strength is a weight factor that gives meaning to related energy flow. The stronger the emotion is, the more powerful is the mentation. However, emotion is physically connected to the degree of strain strength and

---

9 *Fractals for the classroom, Volume 2*, Heinz-Otto Peitgen, Hartmut Jürgens, Dietmar Saupe.

submits an energy flow through a pathway accordingly when released during sleep. In the linear range of stress-strain correlation, emotional intensity is low and emotion is considered as insignificant noises; but on a nonlinear scale emotions are very regulated and clear.

Beyond a certain threshold, when the input becomes too stressful, the formed strains and the resulting attached emotion increase exponentially. Emotions are an alarm meant to keep brain tissues safe from high levels of overstrain. The amygdale, in the middle brain, is the location that emotions emanate from and turn into outputs.

Any on-going energy fractal can have one or more orbits that resonate with one or more substrates in the course of development. Excited substrates become attractors and form the next appearing fractals. The tiny pattern of synchronized firings in a substrate will grow to higher levels of pathways and layers following fractal rules. Any substrate structure is the result of original genetic growth and the deforming consolidations created by overstrain in life. Any overstrain event involves a convolution and a releasing process and, therefore, is attached to general emotions like pain and pleasure. However, depending on the type of configuration, any overstrain represents a specific type of emotion and, when excited, recalls the same emotion. Thus, emotional pain is used to build and grow negative emotions and pleasure is the core used to build and grow positive emotions.

Mentation Alphabet

The number of orbits and fractals that appear by way of firings synchronization is limited in the brain. Physically they are limited by a layer's thickness and a nerve's longitudinal and vertical arrangements. Their quantity can be estimated using their number of extroversion and introversion known products, as with the symbols people use to communicate. Actually, the numbers of letters in different languages does not vary very much. The numbers of basic letters in different languages are not more than 50 and they can normally be compared according to similarities in writing or pronunciation.

Is it a correct deduction that the number of quasi-balanced orbits that can be built into the human language zone that are known up to know is directly related to the letters and symbols in any language? Similarly, the number of emotions in different people is not high. The number of emotion main keywords, as of now, has been listed in most of references at around 50 as well. Therefore, it can be a correct deduction that the number of quasi-balanced orbits which amygdale structure can form has an upper limit of about the same number. The number of frequency bands that human sensory channels are sensitive to can be used to determine the limit of firing orbits and fractals that can be configured in input-output zones.

How many fractals, in family sets, have been found in nature? The brain, as an integrator of inputs, needs to reflect outside facts, to process them and to provide a high-quality understanding of facts in order to protect and allow the growth of the individual. Does this mean that the brain is facilitated to form all of the required number of firing orbits and fractals needed to represent the surrounding facts? Even if it is not completely facilitated to do so, however, it is expected to have followed the same evolution history and to have increased the capability of synapse configurations to detect and picture all surrounding substances, events and circumstances.

The number of orbits and fractal families that have been formulated by humans up to now are not many. By developing the mathematics of fractals in nature, the specialty functions of different brain regions are expected to be known in future. Firing fractals, if known by their configuration, can provide a measuring tool for the configuration of their energy content. The physical property of outside events and facts are translated to energy fractals in the brain and its properties. This means that energy eddy, energy packet or energy wakes contents are all indicators of a variety of pseudo-facts and events. These are the building blocks, made of orbits and fractals, with which the brain deals.

It is known that input and output zones are not so easily distinguished. This fact confirms that fractals drawn from the environment and those projected into the environment are shared and a development in one of them results in a development in the other one.

Mentation Changes

Although different mentations can proceed in parallel, the primary firing fractal is the one that is attracted by focus as an attractor. Any mentation appears when the proceeding mentation is disturbed and disappears; because its memory is not consolidated up to the regeneration period, it is considered as a working memory.

The missing connection between the disappearing and appearing fractals is due to firing gates' configuration changes or, in other words, the elastic behavior of fibers. The route of such a connection could be found using a non-linear dynamic equation that is not solvable in most cases. Therefore, finding a new tool, like pattern animation, is of great importance.

Linear systems have a definite property and that is the reason they can be separated or added to. In the brain, such a function rarely is because the brain is largely dominated by non-linear processes. In non-linear systems, properties can be integrated and combined, but they cannot be added. The brain's non-linearity is accompanied by changes of the fractals in procedural substrates, which those changes act as feedbacks while simultaneously vary the correlation rules of the system. At any time, each substrate in the brain can be excited if it appears that an orbit fits to an initiated firing fractal. The growth of a firing fractal in any specific substrate is directed by influencing attractions. However, any individual orbit growing in a substrate is unstable and changes, potentially on all levels, up to the whole firing fractal growth in the layer level. Therefore, the overstrained location would form at the termination of the growth.

Strange attractions—including short and long term memories, and instinctual and genetic tendencies—determine the specific regulations of fractal patterns built amidst a chaotic condition. Generalization and self-similarity rules for fractals with the same strange attraction determine that the general behaviors of different integrating structures are uniform and convey similar emotions. Thus, any measurement procedure that enables one to categorize a general similarity in pattern defines the behavior category of those patterns. This means that one unique substrate, when it takes part in different mentation processes, can develop different temporary structures that are slightly different from its rest position, and therefore can carry an amount of stress load inside it.

In a given region, if the strange attraction, big or small, can submit sufficient influence on the region, the region keeps its present characteristics. In other words, the fluctuation range of firing orbits in different periods (hourly, daily, monthly, etc.) remains generally the same for a longer time despite changes of inputs during that time; this shows that the location produces the same shape of firing fractal during the chaotic condition of the event of that input. In short, the specialty of the location is formed by the strange attractor.

Each strange attraction in the dynamic system of the brain has a related location and locations have boundary lines between each other. These boundaries have a flexible nature that describes possible fluctuations between attractions and that describes all kinds of structural states that can develop under different circumstances.

If these changes were drawn in a phase space, in any moment, each state of the location gates configuration could be indicated by a point in that. When the configuration of the location is changing, the point moves to a new location. Connecting such points gives a curve. Any influencing variable, which individually can change the configuration, adds a new degree of freedom to it and the long-term behavior of the system becomes more complicated in the event of future location stimulation. This means that, although the specialty of locations remains the same for a while, this develops with the saving of new memories.

The fitting of a suitable fractional pattern to the attraction point could give an idea of the possible pathways an attractor can develop. In a complex pathway, if the numbers of influencing attractors changes, the pathway route would not be predictable and a new phase study would require determining the possible route for the complex pathway.

Change in a pathway results in the variation of related mentation. Considering the above explanation, the missing connection between subsequent mentations can be traced by tracing discontinuities in the phase space drawings. With this concept in mind, firing orbits, or fractals in substrates in a region, will encounter a change in the amplitude and frequency of the oscillations, limited by structural viscoelasticity property. This confirms that the similarity of neurons in a location and, consequently, the configuration of synapses in any location, enforces a limit on related firing gates configuration and size; therefore, this places a limit on possible orbits and fractal shapes in the location, their natural frequency range, their capability to absorb and confine the incoming free energy packets and their specialty.

Symbols and Their Development

Symbols are separate pulses of energy configurations from a pathway into output zone. The kinematics of a configuration reflects the emotion associated with the pulse during this process of energy transmission. In other words, any configuration's motion is accompanied by a stress force that is followed by a strain. The release of the strain is sensed as an emotion. Meanwhile, emotions are expressed by related configurations and motions during their release and are followed by a definite impression and feeling. Emotions decompose into different static-dynamic components and these are all equivalent symbols for that emotion.

In a slow motion scenario, configurations and motions in input zones are mixed by reducing intervals of pulsing between the states of configuration and motion during processing. The state of maximum mixing reflects the emotional content of a mixture. This is made possible when a configuration and its related motion possess the highest frame-time in input-output zones and the least variable frame-time in the outer brain during processing.

Slow motion scenarios occur 1) in input zones, before the initiation of any interaction between incoming forces and pathway configuration; 2) in output zones in which the pathway has absorbed the imposed stresses in its elasticity range and the configuration starts to release itself from strains, thus providing pulses of firing fractals and motions. During the processing period, the interaction between shapes and motions is tremendously high and the temporarily deformed configuration moves like a snake. This happens in the outer layer bridging inputs and outputs. The firing fractals in input and output zones are attracted by the excited saved memories of shapes and events in the hypothalamus of the middle brain; this occurs when the patterns of energy transfer along a processing period are directed by emotion attractors in the amygdale of the middle brain. These attractors themselves are basically formed by need fulfillments in the inner brain and help to push frame-times exponentially high or low.

Accordingly, shapes and properties are reflected as configurations of energy and stresses, and they impose strains on the synapses and neuron fibers. Due to the elastic property of the brain's material, strains are released as confined energy into output zones, reflecting a locations' shapes and motions. In this chain, surrounding continuums (as substances and events) are converted into strains (as emotions) and finally relaxed by internal expression. Then, if the internal expression is strong enough, it is sent as signals to be expressed externally. This cycle occurs during waking time. This process of conveying internal expressions externally became possible with the development of the language area in the human brain.

Although the strains due to stresses in the linear range continue to be relieved, a portion of higher strains, caused by high stresses in the nonlinear range of the correlation, remain and cause the elastic material of the brain to become stiffer and decay in elasticity. By reducing elasticity, brain processing will decline in efficiency and need to be recovered by full relieving the strain during a regeneration period (sleep). The cycle occurring during sleep emanates from remaining strained locations, with the strain released into output zones and presented as an internal expression by exciting the firing fractals of saved memories through the pathway.

The higher the emotion load of a mentation, the higher the expression length; the shorter the expression statement and the less ordered the statement and less understood. A dream's narrative is made of scenes, characters and the characters' moves; and all the above specifications apply to it.

Input and output zones are common in the related zone. However, output zones receive more abstract shapes and motions because of their more complex interactions and developed attractors, but these shapes and motions will be shared with input zones when expressed externally. This way, symbols increase in complexity. During waking time, complexity provides depth to meaning and communication. In dreams, pathways are random and short, and products are expressed by random and with presentation of symbols which reflect a definite emotion; emotion is the way that the free energy releases and therefore in dreams, emotions are felt very strong.

In short, inputs impose strains and the energy content of strains are released as outputs and memories.

Mentation Continuity

Brain activities are basically made of synchronized firing patterns. A continuous flow of these patterns carries the amount of free energy needed to activate motor outputs to express or carry out an associated action accordingly. An image of such a pattern will be saved as a memory with a working, short or long-term duration. Past saved patterns play an important role in the formation of present and future patterns of mentations.

Any flow is initiated by an input, either internal or external. All patterns flowing at the same time can be tied together in a complex pathway, if the growing firing fractal shape is preserved. Otherwise, if an active pattern disturbs a dominant pattern, there will be a change to the pathway or a disordered output will be created. An introduction to mentation continuity and termination was given in "Mentation Population Growth," in volume I of this book series. Strange attractors, such as the configurations that recall the previous saved memories and genetic patterns, are the basis for the creation of new fractals.

Continuity, in energy flows and segregated fractals, can be detected by observing the sequence of brain activity outputs, as observed in actions or expressions. When brain activities are presented in a questioning output—when the output still needs some feedback input to complete the expression—there are several linked pathways that are governed by a general pathway. Such an integration of complementary inputs that finalize a mentation is continued in the frontal lobe.

Separate pathways may process different mentations at the same time. Lucid thoughts and dreams are examples of this. People with the capability of focusing on two or more events at the same time are another example. However, one always remains as the dominating pathway. In general, pathways running at the same time are in the same family of pathways and reinforce each other, although they may look different.

Modulated pathways are those that are filtered so that they admit only a narrow band of frequency and are intended to produce one unique result; these are formed by training. Those mentations that are modulated to accommodate a wide range of linked pathways and complimentary inputs reflect more highly examined mentations and more reliable truths.

Mentations terminate in an emotion, solution or conclusion, which is followed by an action or expression, but they can also be conditionally ended, to be confirmed by facts and experiments. Conditionally ended mentations are partially completed and they cause the running pathway to be excited, if they are not disturbed by other inputs and potential feedback inputs. In that case, they would naturally iterate a closed pathway with a feedback error that regresses to a minimum.

The frontal lobe has the ability to receive wide-spread pathway products and to modulate them to create one unique comprehensive ending. However, without continuation of the pathways over the frontal lobe, pathways jump from one branch of pathway to another and are never integrated in a unique ending, because of changes in the strength of the involved attractors. Such big changes are dumped in the frontal lobe.

Sensing is prior to and initiates energy fractal growth. Energy fractals recall the relevant images and reflect the attached feelings. All sensing, as inputs and outputs, is converted into an energy flow in a specific configuration that interacts with the travelling structural configurations of synapse and fiber strings in the way. The interface between energy media and consciousness is their attached overstrains and emotions. Emotions associated with an increase in overstrain are painful and negative. Emotions associated with the release of overstrains are joyful and positive. Physically pain and pleasure are the general cores of any complex emotions that develop in an emotional tree of further developed emotions in an evolutionary order. The overstrains' thresholds, as experienced before the new experience of a stressful event, send the related confined energy that is released to a special structure in the amygdale, in the middle brain, to be processed and sent out as related feelings and/or actions.

High emotional brain activities are more informative in reflecting the circumstances of a given situation. Low emotional brain activities are more focused on describing subjects and events in details. Consequently, feeling is a qualitative measure reflecting circumstances, automatic thought is a semi-quantitative measure reflecting typical subjects and events and intentional thinking is a quantitative acknowledgement reflecting specific subjects and events.

Onset of a Mentation Appearance

The fractals that appear in the brain, as a self-organizing system, follow a general mathematical model; based on this model, firing or energy particle patterns can be traced based on the related amplitude and natural frequency while considering the time and the position of fractals in different locations of the brain outer layer. The mathematical model of self-organizing appearance consists of a driving term for energy transfer; limits to the pathway boundaries and energy diffusivity or dissipation terms through the pathway[10]. The apparent pattern can be recreated by different shapes of spirals, waves or topological defects. The shape of a pattern depends on the dimensions, structure and physical properties of that layer on which the pattern is configured.

---

10 Christoph Schiller, *Motion Mountain, volume 1*, p. 323

Synapses and a Firing Fractal Growth Over the Them

To predict a subconscious brain activity, a picture of memory arrangement in a tree-shape configuration can be drawn. This figure would have leaves and connections representing subjects and events in the middle brain and branches and sub-branches in the outer brain, where core attractors in different locations, including in the inner brain, shape the curvatures of the branches as layers of energy flow. The branches in the brain's outer layer determine the routes taken by energy layers of synchronized firings waves. The branch's movements are governed by functions that grow the leaves and connections. At the same time, leaves and connections absorb nutrition to feed the tree. There are groups of fixed leaves and connections serving as substances and properties composing the stage; flexible leaves and connections represent players and their behaviors. Each group builds a memory when excited by an input. The sub-branches' movements are limited to the connecting main branches' movements and all are connected to the tree column all together make individuality and personality.

The memories are laid out in a specific manner in different areas so that the family memories that recall similar emotions are saved in the same location. When a new memory is in a building stage, it will grow on the top of a past memory. Therefore, the grown structure determines the process of categorization. When a pathway is excited, the building memories are excited and energy trapped during the flow of energy through the pathway can potentially grow that structure. A mentation consists of the newly added and forming fine configurations and the excitation of the pieces of memories through that. All story making and logical reasoning involve recalling the previous memories as experienced and as trained for, in addition to newly added values such as feelings and conclusions. The relationship between the connected branches and groups of them are defined in the language of symbols and metaphors.

The leaves and branches are exposed to input winds and the felt emotions indicate the highest acceptable strains that are due to wind stresses. But they absorb the input momentum temporarily, before extreme emotional alarm, either by relieving themselves when the stressing source is removed, or by relieving and reconfiguring themselves when the inputs are blocked during sleep.

The strained part of synapse branches and neuron fibers relieve and reconfigure themselves by distributing their confined energy within the group and breaking the full emotion load into smaller loads. However, during the whole process they follow the fractal rule of self-similarity production. The emotional mood, as a governing function, remains the same, although it has slight differences of emotional contents in the same category. Each orbit in the fractal is a small bundle of synapses split into smaller bundles for all involved layers.

Each layer of energy orbits that has amplitude, frequency and motional spin represents a player; each motion, then, represents a behavior, and the stage would consist of miscellaneous filling substances and properties, related or not to other components involved. To construct a fractal, orbits retain the same structural configuration as the key emotion, although emotion intensity differs by a scale.

Energy Flow Through a Growing Firing Fractal

The extent of each mentation (a unit of brain activity) depends on the stability of the growing fractal and the memories that the grown fractal contains. A mentation is terminated when the growing firing energy fractal is terminated, when all firings within a fractal are synchronized and by the synchrony disturbance. In waking, the fractal growth is limited by synapse configurations, neuron types and neuron fiber structures in the location of a fractal, as well as temporary deformations that develop due to strains. In sleep, fractal growth is limited by the amount of confined energy caused by the strains built up in the waking hours. In short, a brain activity consists of the flow of configuration energy through a pathway that covers local synapse structures as memories.

In waking, sensory and internal inputs initiate flows of energy through fractal growth; these flows' directions are guided by attractors. In this process some strains are accumulated, which decays processing efficiency. In sleep, strains initiate energy flows through a fractal growth; these flows' directions are guided by memories and other attractors. Attractors, by directing energy flow, determine the concentration and focus of a pathway, either subconsciously or consciously.

Transfer of Configuration Energy Through the Outer Layer

An energy balance around a substrate can be integrated for a pathway and following present a general description of it in mathematical expression. ε is an abbreviation for strain and τ is an abbreviation for stress in the formula:

$\square\varepsilon/\square t = f(\tau)$; then: $\tau = \varphi(\varepsilon_{d1}^a, \ \varepsilon_{di}^i, ..., \varepsilon_{dn}^n,)$ where $\varepsilon_{d1}^a, ...,$ ε_{dn}^n are strains as a function coefficient in the interaction process. Brain waves are an indication of oscillations during the transfer.'s develop in the amygdale day by day and decompose into the static and dynamic elements associated to them. New patterns are created in these layers to balance the configuration energy throughout them. The balancing process consists of two stages, consolidation and uniformity, explaining physically the building of a new memory and a dream scenario, respectively.

- Individual static and dynamic information is entered from sensory channels into the input zone.

- ε's, in linear range of ε-τ, form and are removed by ending the τ inputting, but remain for high tension τ's in nonlinear ranges of the relationship. They will be stored in the same family pattern locations temporarily, as confined energy in the location.

- The pathway of Є-transfer is influenced by attractors in other areas and creates attractors when the transfer is ended; the created attractors are stored in the amygdale as new emotions.

- Temporary Є's will be consolidated during NREM stages of sleep and would be distributed uniformly during REM stages of sleep, as emotions in the amygdale and as static and dynamic memories in the hippocampus.

- Through configuration energy transfers, information is transferred through synapses and images through the layer pathways.

- The input information initiating a brain process in the input zone and the integration of processed products in the output zone are shared and are not distinguishable.

Stability of State

The classical curvature elements, which are used in schematics showing phase spaces in a plan for energy distributions, are as follows:

A system defined by { $\dot{x} = x$, $y = \dot{y}$ } is a limited system and unstable at point (0,0)	
A system defined by { $\dot{x} = x$, $y = \dot{y}$ } is limited and stable at (0,0).	

A system defined by { $\dot{x} = x$, $\dot{y} = 2y$ } is unlimited and unstable	
A system defined by { $\dot{x} = x$, $\dot{y} = 2y$ } is unlimited and stable.	

Following states are indicated by curves:

a) Unstable (saddle point)	c) Stable	e) Unstable
b) Unstable	d) Stable	f) Stable (drag point)

Energy Distribution Index

In fluid dynamics, Reynolds' number is a familiar index for estimating the state condition of a fluid:

- If it is less than 2000, the flow characteristic is laminar or steady.

- If it increases to above 4100, then flow state changes to first transitional condition.

- When it increases to more than 4100, turbulent flow exists. As the *index* increases, the chaotic characteristic will increase as well.

Reynolds' number is a non-dimensional and is measured by the division of kinematics-to-potential energies or by momentum-to-inertial magnitudes, both of which reduces to products of *fluid density, average velocity* and *nominal pathway diameter* per *medium viscosity*.

Laminar flow occurs when the energy streams or the pathway curvature of particles are clearly distinguished; in such a condition the pathway is determinative and that can be described by a definite equation. A transitional flow displays some chaotic behavior, but the average route of particles are described by a linear equation for this type of pathway. When the Reynolds index is increased, turbulence increases, which means the non-linearity of pathway particles increases. Therefore, it can generally be said that, if an index can give a non-linearity description and an increase of the index is proportional to the non-linearity of the changes, then variation of the continuum, from a steady state to full turbulency, can be demonstrated as measured against that index.

For example, if the increasing rate of a population is assumed as an *index*, used to foresee the future population of a given mentation, the formula model is $x2=r*x1*(1-x1)$, when $x2$ is the number of the period following $x1$. It can be researched if the repeating mentation, as a mood, follows the same model. For the examination of each following period, the previous x2 should be substituted for x1. Then, by plotting the population, the curve falls into sub-branches and it can be seen that:

- When the *index* (r) is around 2.7, population remain the same.

- When the *index* (r) slightly increases, population increases linearly.

- When the *index* (r) is around 3, the curve will be divided into two branches; this means the population can find either of those alternative population numbers.

By increasing *index* (r) more and more, the resulting splitting multiplication can produce different possibilities of the estimation of the population figure. In any period, figures are presented with more options as compared to previous instances. The branching of more options continues by increasing in "r" and accordingly the increases in options that a population can find increases. In other words, the state of the mind is not predictable. These dividing turn points are called bifurcation points, which are further solutions to the above mentioned formula.

Other kinds of *index*es are:

- Entropy to time changes

- Kinematic per potential energies

- Momentum per inertial changes

- Oscillation power changes

Activated Substrate and Region States

Any pathway state is an arrangement or combination of the activated sub-states in the same time and on the same route.. The degree of stability in each state depends on the number of routes that terminate to other states of arrangement or combination; furthermore, the life cycle or duration of each state is dependent on its degree of stability.

A continuum, which links a configuration and a motion, can be considered as a complex of sub-continuums made of finer configurations and their motions. By such a description, a pathway can be indicated by a matrix of deforming substrates producing a free flow. A continuum depending on the location and momentum distribution has a content of potential and kinetic energies. Also, each has 2n-dimensions (n geometrical axis and n pulsation modes), the former of which defines localization and the latter of which expresses the motional properties of the continuum.

In term of firings, no location in the brain has fully synchronized firings:

1. A free spot firing has an ideal distribution of amplitude-frequencies if related synapses are not influenced by neighbor effects of making electromagnetic fields in case of synchronizing. Neuron firings will find synchronization under fine electromagnetic field or stress field progress along a pathway. This is the mechanism for the synchronization proceeding along a route during a mentation, which makes the pathway and has been described in volume 1 of this book series.

2. Practically, a group of substrate firings by energy transfer will find a proper distribution of amplitude-frequencies; the peak frequencies in that distribution curve indicate the localization property of an energy flow. This localization concept has no relation to the brain's co-ordination point.

3. The average frequencies of involved substrates' amplitudes show motional properties while the low frequencies, of varying amplitudes, show entropy property.

4. Finally, the low frequencies of very large amplitudes indicate the field-generated properties of that continuum.

Each state of a firings' continuum has the same probability of existence as others, except that the field induction of the neighboring continuums introduces definite patterns, which causes the distribution to deviate from a normal distribution. The life time of each state of a firings' continuum is based on the continuum equivalency degree.

A continuum of firings can be shown as a spatial curvature as an orbit or fractal. A firings' distribution has an equivalency degree that continuously alternates between a minimum and a maximum level of frequency. The minimum is related to rest position and the maximum is connected to being physically the under highest possible tensile stress. So any continuum passes through different ranges of frequencies, including peak amplitudes, a mixture of average amplitudes, a mixture of large-to-small amplitudes, and a mixture of very large-to-small amplitudes. Accordingly, a continuum shows: localization, regulated movement, distorted movement and field influence properties. A continuum with a high frequency of same-sized amplitudes is a hyper-synchronized substrate for the firings. Ideally, such a state presents an image. Obviously, a location with a more flat distribution is less synchronized for firings and practically will not be involved in the pathway. The locations with the least firing synchronization make the boundaries of the pathway of the energy transfer and do not take part in the ongoing brain activity.

Defining the Free Energy Flow Attached to a Brain Activity

Firings, chemical changes, and electromagnetic field packets (on the micro scale), as well as energy packets, momentum transfer, and change of fibers configurations (on the macro scale), are the dynamic parameters that play the main role in the brain by running a mental activity (a mentation) attached to a free energy flow.

In the other side, if we were able to monitor all of these changes in a given period of time and at a given location, we would observe a dynamic flow of eddies which, when settled, would cause some changes in the geometry and the chemical and firing properties of the location. If we look at:

1) folds as rigid sharp flow discontinuities,

2) attractors as a semi-flexible limits, and

3) momentum eddies as energy lumps,

then any initiation by sensory or internal inputs will motivate a flow of energy eddies as described by the above conditions, which terminate in the output zone. The standing configuration, as a wake, is a performed mentation and a working memory. At the beginning of any mentation, the energy flowing with random velocity and fluctuations, and in different directions, is turbulent and chaotic. Due to the thinness of the layer, the flow moving in a vertical direction is ignored.

Energy eddies and packets are continuously and constantly interacting with one another (with negligible time lags) so that they cannot be recognized as individual units in normal timeframes. The eddy flowing in the pathway initially has a wide range of sizes and is therefore most chaotic. The size of eddies is proportional to the square of their distances from big plasticized curvatures, like folds. In continuation of their transmitting, they move in a more steady state, and finally they find a creeping flow shape of more complicated and smaller-sized eddies. Length as an indication of eddy size is proportional to the distance from the folds.

Eddy viscosity, which is a strong function of position and time, is a product of energy density and momentum diffusivity in the pathway; momentum diffusivity is a product of the substrates surface, which host eddies and a second-order derivative of the stream function along a direction perpendicular to the surface (the energy flux). Therefore, eddy viscosity is expected to be very high in the outer layer's back side and negligible in the layer's front side. This provides the conclusion that folds concentration is higher in the back of the brain and lower in the frontal lobe. However, this is an assumption that needs to be confirmed experimentally.

Kinematic energy or stream function is as important as potential energy in the back side, while high potential distribution in boundaries determines the pathway. These energies are known as pattern geometry energy. Big eddies appear as form drags, and we refer to them as orbits in other places. When eddies and packets, via interactions in a pathway, grow into a bigger size, they form more static drags; the increase in size, according to the conservation law, will lead to an increase in inertia and a reduction of kinetic energy. A mass of eddies shape a wake made of drags with low-travelling velocity. The flow lines inside any of these energy units, because of their clear streamlines, are laminar.

A local and timely distribution of eddies forms a dynamic pattern of spatial rotating eddies which, in turn, configure drags. Drags move forward in the pathway and create wake patterns. Momentum convection flux, caused by eddies, drags, and wakes, is a product of related mean stream-function differentials and energy density. It is also the negative product of eddy viscosity and the second order differential of stream function along the direction of the free energy flow. Eddy viscosity itself is a product of density and eddy diffusion. Therefore, it is concluded that eddy diffusion is a product of the eddy substrates surface and the energy flow flux rate.

Conceptually, if eddy configuration can be thought of as the alphabet, drags can be assumed to be words; words, then, would virtually build wakes, which are represented by sentences, and a rational combination of sentences, when put together, produces a mentation.

The nature of an area *specialty*, to some extent, shows what kind of pattern the energy packets will develop into, when flowing through that area. The pattern presenting the local function indicates the substrate size, distribution during transfer and the physical property of the location. The finer the energy particles, the more quickly they move. The occurrence of larger sized energy particles increases the absorption of the transferring energy because confined stress increases the potential energy of locations. This causes more locations to initiate an internal input when the source is removed. Furthermore, when the location is under process, its oscillation amplitude is stronger.

It can be concluded that developed branches in a mentation are as condensed as is a complex in the outer layer's front side. The condensation process would occur with the reverse transfer of energy from the layer level to the synapse level, providing the same shape fractals. The fractal reconsolidates the location with a new configuration suitable to initiate that fractal in the future and to facilitate new logic functioning and reasoning in the location. The new arrangement is fixed in the location based on physical properties and their possible range of configuration fluctuations.

In summary, the flow of mentation through the outer layer creates chaotic, transmittal, laminar states and, finally, structures the location.

A Substrate or a Region State

Any state of a substrate fiber configuration caused by confined energy has a static character, which can be different from its rest state. Similarly, it has a dynamic character that can be observed in a substrate's oscillating behavior. Any static arrangement in a substrate is a combination of its sub-substrate arrangements. Therefore, any substrate structure has a stability degree based on the number of routes terminated in other sub-substrate arrangements. In other words, any synapses-fibers configuration's life-duration could be foreseen based on a stability analysis of their smaller group arrangements.

If a configuration could take place amidst a higher degree of rigidity, this would create a more complicated structure. On the other hand, any potential for change between two different states is a potential for an interaction in the location and reveals how dynamic the arrangement of sub-substrates is. Any potential interaction in a location has an internal equivalency degree (kinematic energy per potential energy ratio) related to the rearrangement possibility within a substrate, as well as an external equivalency degree, when an interaction may occur with other substrates. The stability of an interaction configuration can be predicted based on the combination of the sub-substrate's equivalency degrees. Any interaction can be configured in a schematic way of structured–flow (like an orbit), regularized-flow (like a fractal), transmittal flow (like a distorting fractal) or turbulent flow, based on the combination of the equivalency degrees of interacting elements.

The continuum in or function of a location is the integration of structure and interaction, or a state with both static and dynamic characters limiting structural changes. Structural changes in a substrate within a region are followed by changes in interactions that build a specific arrangement according to the dominating continuum. Therefore, the substrate function is a combination of "equilibrium" and "non-equilibrium" conditions of a substrate's fibers configuration. Non- equilibrium condition is governed by the rules of increasing entropy and the irreversibility of change in configuration. Generally, any substrate is not in a rest position and, due to incoming information or noises it maintains a distance from the equilibrium condition. The distance from equivalency shows the load of stress or confined energy inside the substrate. While in an equilibrium condition, in which the substrate is not excited, the production of entropy is semi-constant, revealing that the condition is close-to or at-equilibrium or rest condition. When distance-from-equilibrium, or the rate of entropy production, increases to beyond a bifurcation point, the fractal configuration of the firing gates in the substrate (beyond a threshold of the entropy production rate when self-similar production of synchronized firings will stop) disappears and a newly ordered structure will be created.

The entropy production rate is indicative of the distance from equilibrium. This means that energy state changes from ordered structures to steady, transmittal, and turbulent transferring states and that the energy fractal changes from a clear configuration to a full disappearance. This also indicates a threshold for a new or more complicated fractal appearance. This periodic change is repeated and repeated again in a densely branched way. Most of the produced entropies are absorbed by the deforming of the rest configurations, up to a point that the physical structure of region does not allow any further stress absorption. This means that entropy production should be decreased to zero and will be develop a negative sign in the final strain releasing stage of the regeneration period.

Rules of Brain Functioning

Because any mentation is attached to an energy transfer, similar rules apply to it that apply to any other transfer media; thus, the rate of transfer is equal to driving force per resistance against the flow. The energy transfer imposes momentum forces on pathway surfaces. In any pathway, the building substrates have a different energy density and fluctuating velocities, relative to the energy layers' configuration on the substrates. Therefore, momentum transfer along any pathway (z-direction) can be expressed by the product of momentum diffusivity factor (v) and energy flux ($v_x \int$) $gradint$:,

$$\tau_{zx} = -v \frac{d(v_x \int)}{dz} \text{ where:}$$

v_x *is the energy pocket transfer rate and $\int$ is energy density.* Although the momentum diffusivity factor (v) is generally non-linear and time-dependent for a viscoelastic media like the brain, for low stresses it is almost linear. In other words, low stresses normally come and go by being absorbed and released with the same amount of momentum, which facilitates the return of fiber positions to almost their origin points and therefore leaves slight or no traces of activity memory.

By creating a general momentum balance between absorption, interaction and desorption around a substrate or a pathway, the following formula would describe the outcome in any dimension:

Generation rate = accumulation rate – (output-input rate)

$$R = \frac{\partial(v_z)}{\partial t} - v\,\frac{\partial^2(v_z \int)}{\partial z^2}$$

R is momentum generation rate, which is positive during waking time and negative during sleep.

Balancing can be performed for momentum-resulting forces as well:

Inertia forces = input forces - viscous forces

Inertia forces are due to energy density; input forces work as drives and viscous forces play as resistance to transfer by deformation. The ratio of forces defines the characteristics of flow, which can be shown by two indexes which are Reynolds and Euler indexes.

The instability degree of the energy flow is determined by the ratio of the kinetic or inertial forces ($\int \bar{v}^2$) to the viscous forces ($\mu \bar{v}/\Gamma$): Reynolds Index= (inertia force)/ (viscous force)

The sensitivity of the substrate or pathway dynamics to input is determined by the ratio of input force to inertia force: Euler Index= (input force) / (inertia force)

Geometrical Index = (energy packet mean free path length) / (substrate characteristic length)

The geometrical index is a powered product of the function of the Reynolds Index and the Euler Index. The geometrical index is a typical combination of the two mentioned indexes: Geometrical index *(Renolds Number)^a*= (Euler Number)^b*. The reason for expecting an exponential relation for the indexes returns back to their combinational influences which has contradiction with a linear correlation. The exact relation is to be found by using the experimental data in specific designed experiments. Consequently, geometrical sizes can be determined by finding the unknown numbers of a and b.

If viscous forces are high compared to inertia forces, then the momentum diffusivity factor is linear and there is no accumulation, which means the transfer will not be saved as memory and will not have an impact on the layer significant enough to require a regeneration process. But if the inertia force is strong, then the energy transfer is turbulent and the momentum diffusivity factor is nonlinear; in this case, there would be an impact on the layer and accumulated energy as confined energy would leave a trace of the transfer as its working memory to be fixed in the regeneration period.

In this case, the confined energy parameters are proportional to the frequency and ratio of amplitude changes per substrate characteristic length for each substrate on the pathway. According to the above explanations, roles of recalling and saving in each activation process are as follows:

Rule 1: Substrates or pathways that are geometrically similar are dynamically similar, if the parameters representing ratios of imposing momentums are equal. This means that similar geometrical substrates, when they handle an energy packet flow and their Reynolds and Euler numbers are the same on that location and time, will reactivate the same memory (either of substances, or names or functions). Self-similarity means geometrical similarity.

Rule 2: Dynamic pathways create a similar geometry of synapse configuration in the local or in a far away substrate, if they are energetic enough to transfer or to make a resonance in their flexible structure. This means similar geometry will be built in a substrate; if the Reynolds and Euler numbers are the same at the time, then they build a memory of a pathway pattern, which represents a specific input situation.

Rule 3: Reynolds and Euler numbers are correlated with each other and attracting substrates (memory substrates) and the functioning pathway have a scale-up / scale-down relation in the course of the recalling or memory-saving processes.

Frequency and Natural Frequency

Frequency is used to measure a single event happening during a unit of time. Natural frequency is a combination of integrated events in a definite structure and therefore is a combination of frequencies. In the case of momentum transfer, the parameters which generally determine a structure are dimensionless indexes which are unique for similar structures, regardless of their size.

Inside energy packets, eddies rotate in definite stream layers and there are separate rotating events in separate layers. The natural frequency of energy packet or pathway wake is a combination of separate frequencies, each associated with different rotating lengths, energy density and viscous property. The weights that relatively measure frequencies in a structure are combining factors. Therefore, natural frequency is a function of Reynolds number (in respect to dynamic) and Euler number (in respect to input) and characteristic length ratio (in respect to geometry) in any substrate. If all of these indexes will be the same for two substrates, or for a substrate and a pathway, or for two pathways, they behave similarly.

Recalling and Saving Rules

According to the above, rules for recalling and saving in two similarly activated substrates or pathways are as follows:

Rule 1: Substrates or pathways which are geometrically similar are dynamically similar, provided initiating inputs are similar. This means similar geometrical substrates, when similar energy packets flow through them, show similar dynamic behaviors. In other words, they reactivate the same memory (either of substances, or events or functions).

Rule 2: Dynamic pathways create a similar geometry of synapse configuration in the local or in a far away substrate, if they are energetic enough to transfer or to make a resonance in their flexible structure. This means similar geometry will be built in a substrate; if the Reynolds and Euler numbers are the same at the time, then they build a memory of a pathway pattern, which represents a specific input situation. Self-similarity is geometrical similarity is one of three requested parameters.

Rule 3: Reynolds, Euler and characteristic length ratio numbers are exponentially correlated with each other and attracting substrate (memory substrates) and the functioning pathway can be a scale-up or scale-down of each other, in the course of memorizing or saving memory processes.

Mirror Pathways

Pathways made by so-called mirror neurons are the pathways that transmit a free energy flow goes under high deformation process and the same or a similar input is repeated as the performed motor act as well as will be consolidated as an specialty. Such a behavior is clearly seen in numerous animals as well as humans, especially children, whose brains are not yet as folded as those of adults. This behavior makes many areas of learning possible. Children like very much to watch adults walking when it is the time to walk, or to watch others speaking in order to learn how to produce the same sound. Language learning is one of most important areas of learning in the early years of life. Such a behavior is famous for being observed in monkey and the imitating of words by some birds.

Tracing a pathway

Discrimination between inputs depends on the strength of the input pulses compared to the activation energy of the locations. All kinds of energies of chemical, electromagnetic and nutrition are components that determine neuron firings situation including synchrony, density and the geometry they are arranged in and for simplicity is referred to configuration energy here. For example, stress and a chemical (norepinephrine) concentration are closely related in any substrate. An increase of norepinephrine causes a higher flow of blood, which is fed to the strained location to do a fight or flight response aiming to reduce stress. In addition to that, norepinephrine works as a neurotransmitter, forwarding nerve impulses and distributing stress to neighboring substrates. Fight or flight is a response to incoming stress and a way to release it.

However, physically stress is followed by strains on tissues and consequently on firing gates spatial distribution. It means norepinephrine component is directly related to the configuration.

> When we experience excessive stress—whether from internal worry or external circumstance—a bodily reaction is triggered, called the "fight or flight" response. Originally discovered by the great Harvard physiologist Walter Cannon, this response is hard-wired into our brains and represents a genetic wisdom designed to protect us from bodily harm. This response actually

corresponds to an area of our brain called the hypo-thalamus, which—when stimulated—initiates a se-quence of nerve cell firing and chemical release that prepares our body for running or fighting[11].

Furthermore a pathway can be traced with the same chemical component as the index of stress intensity:

The shift from one input to another one (either sen-sory or internal) is influenced by the over-activation or under-activation of the attention center (Reticular Activating System) caused by change in level of Norepinephrine in the related location. The Reticular Activating System connects the spinal cord, where it receives information from sensory inputs, to the mid-brain. The Norepinephrine level increases follow-ing a stress introduced either from the spinal cord to the outer layer or the spinal cord to the frontal lobe, initiating an energy transfer. This process happens through the resonance between sensory inputs and initiated pathway substrates. The same phenomena should happen when the wake energy of a pathway is sent out by motor outputs to perform actions or ex-pressions in the environment[12].

In conclusion, the tuning of substrates in a pathway with in-comings and outgoings happens because of norepinephrine, one of chemicals which are an index of stress intensity in the loca-tions. According to Douglas Cowan, the "Reticular Activating System is a very complex collection of neurons that serve as a point of convergence for signals from the external world and from the interior environment.[13]"

---

11 Neil F. Neimark, M.D., retrieved from: www.thebodysoulconnection.com
12 Ibid.
13 Cowen, Douglas, "The Reticular Activating System, and its Role in Attention Deficit Hyperactivity Disorder, retrieved from www.newideas.net/adhd/neurology

Distance from Equivalency

The ratio of kinematical energy per potential energy in any location of a pathway is an indication of "equivalency degree" or "distance from equivalency." Kinematic energy is related to the speed of changes in frequency; it describes when potential energy is proportional to amplitude. Therefore, this ratio, as an index, can be expressed by a complex function of frequency and amplitude. An infinitely small change from an equivalency position will cause a big change in the function and consequently an increase in the dispersion of energy packets all around.

Therefore, the maximization and minimization frequency of amplitudes dispersion is a fluctuation function around an equivalency axis. The ratio of kinematic energy to potential energy is an indication of distance from the equivalency state:

1) If the ratio is too low, the maximization and minimization of the frequencies follow an order. Clear configurations happen in this range and amplitude-frequency distribution is most pick-wised, meaning a definite small range of amplitude will have a significantly higher frequency, which distinguishes this range of amplitudes from other ranges.

2) If the ratio is too high, the maximization and minimization of the frequencies develop irregular configurations and the amplitudes have a wide frequency distribution. This means the verity of amplitudes is very high and the motion is irregular.

3) If the ratio is between those extremes, more or less semi-regular motion can occur. A regularized to semi-turbulent motion with amplitude-frequency distributions more or less in the normal range would exist in this range.

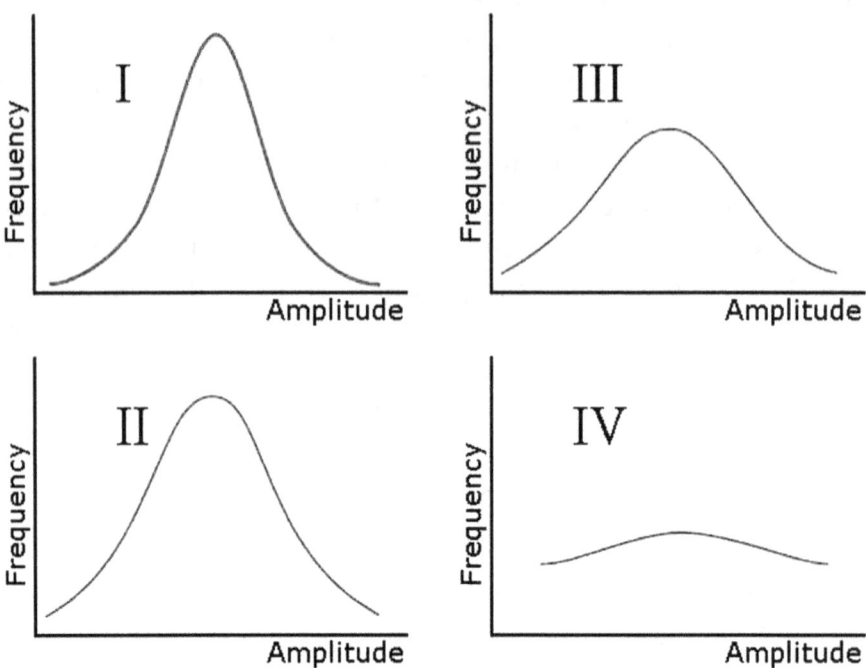

Accordingly, there are spatial changes in amplitude and longitudinal changes along a pathway. Depending on the distance from equivalency: when travelling velocity of changes is low, the formation of patterns occurs; when the travelling velocity of changes is high, the patterns degrade. However, the process is intervallic and with further increases from equivalency, patterns will find higher complexity. In other words, interval increases of irregularity will be interrupted by the forming of higher orders.

To define energy transfer status along a pathway:

a) any longitudinal energy transfer over a potential pathway makes the energy pockets configure in two different types of patterns:

1) Areas with closed patterns. Such an area is a stable and absorbs energy as a temporary strain.

2) Areas with open patterns. Energy pockets flow in such an area is discontinuous manner. These areas are unstable, releasing easily the received energy.

Pulse-wise fluctuations happen in areas with a closed loop pattern and smooth fluctuations in areas with an open loop pattern. The patterns' frequency and amplitude have a definite correlation to each other following the area specialty.

Generally, if a pathway would be considered as an energy plan, the energy transfer follows the following rules:

- pick points of energy layers are unstable substrates;

- substrates around temporary equivalencies are running away locations;

- substrates located in energy layer valleys are semi-stable curves;

- substrates located in plain layers are as-remain points.

b) Crossing energy transferring layers, if any, are the locations of stable amplitude and with steady fluctuation frequency. Further to the state of synchrony, the oscillation frequency works as a strength factor for the links between substrates in a pathway. Frequencies and amplitudes for a substrate determine its natural frequency which should be in harmony through the pathway, giving a specific energy receiving-dispatching character to the pathway substrates.

Energy Transfer and Attached Mentation

Dreams represent a pattern of strain releasing of the locations in the pathways with daily high-stressful inputs. If emotion can be physically defined as the pattern of strains, any symbol in the dream and the whole narrative of that should convey such a pattern. Such patterns depend on the person's experiences of the very stressful events in life.

If symbols are a reflex of static imaginations, metaphors and similes of any type revive patterns of feelings or thoughts as dynamic imaginations. The output zone deals with all kinds of imagination and awareness of them is by the related emotion; this is where along related pathways, only configuration energy transfer and interactions within substrates, as fully physical phenomena, occurs.

Similar explanation can be given for waking brain activities considering that energy patterns made of synchronized firings in a definite gates configuration are attached to sensory inputs as well as motor output through the Reticular Activating System as briefly discussed in the "Tracing a Pathway" in this chapter.

Transfer of Free Energy and Pathway Formation

Sensations in input zones, and feelings or thoughts products in output zone, are all emotional and self-informative. They are expressible if contain sufficient potential to be processed in the frontal lobe. What links a sensation to a feeling or thought is the relevant emotion as a strain-mark of the stress flow in the pathway which connects the sensation to the output. Pain is one of basic emotions attached to a pathway which is initiated with a stressful input. It is an alarm and is meant that some brain tissues are being overstrained and commend a remedy to be adopted to reduce the stressful input. A saved memory of geometry of the tissues under overstrain is the emotion attached to the causing input.

Functions that maintain a feeling flow are emotions that are saved and located in the middle brain. Animals with a developed middle brain, like birds and mammals, similarly recall emotions that urge them to express individualized behaviors. The middle brain—as opposed to the inner brain, which preserves non-trained self-protection with reflexive responses—spurs action in a more complex way and with a time lag. It covers all circumstantial needs, like having food, sex, social connections, emotional relationships, group motivation and play in the way learned or trained during the life.

The instinct of preservation evolves into a more complex tendency that combines preservation with self-advancing. Here, the complexity in behavior is determined by a sense of self. The self is the interface between external circumstances and emotional alarms, and it has a target of needs and desires satisfaction. In the outer layer, especially in the frontal lobe, pathways are capable of producing optional products and to choose between them. In middle brain there are only two options of agreeable and not-agreeable responses to inputs; and in the inner brain, there is only one way of reflexive response. Structurally, the type of neurons—uni-polar, bipolar or multi-polar cellular structure—may explain the types of products in different layers of the brain. Generally, neurons, including motor types, are categorized as single (unipolar), dual (bipolar), or mirror (multipolar).

The amygdale structure in the middle brain supports the behavioral expression of emotion as the accompanied by-product to the rational motor action outputs. The following are some strong theories can help us to conceptualize the formation of a pathway during a mentation:

1) According to the Cannon-Bard Theory, physiological reactions and emotional experiences occur simultaneously[14].

   This confirms the theory that emotion is the energy release of overstrained tissues meant to protect them from injury during a relevant brain activity; this happens everywhere in brain, but would recall the similar saved emotion substrates in the middle layer.

---

14 "Limbic System: The Center of Emotions," by Júlio Rocha do Amaral & Jorge Martins de Oliveira, http://www.healing-arts.org/n-r-limbic.html.

2) According to a James Papez study, emotion is not a function of any specific brain region or center. However, the connectivity of expressing motor-neurons occurs through structures like the amygdale and specific areas, because their specialty gives a specific shape and content to emotion. Similarly, in the kinematic model, it is assumed the pathways are located over the outer layer and all other resonating parts are located in other brain layers and parts, as attractors for the main pathway. Overstrains would mainly occur in the outer layer and are considered as an emotion when resonate an emotion substrate in the amygdale. Emotions alarm the necessity to relief from overstrains anywhere in the brain, and alarm communication is with the amygdale, in the middle brain. The remedy action either free expression relieving or action will be followed by emotion awareness.

3) The regulation of emotional behavior is achieved through the storing of extra energy in the middle brain and by relieving for a part of the strain into the multi-polar neuron structure of the outer brain, which distributes the excess energies by illusions or self-talking.

4) If individual memories and related emotion experienced are saved in the middle brain, the same-family memories make the functional memories in the outer brain. It seems that the middle brain shows a kind of semi-plasticity to deformations to save individual events and circumstantial memories as opposed to the outer brain which show higher elasticity behavior toward stress flow. In the same way, the inner brain is very rigid against reconfigurations, where it conductivity rate is high and transfer the signals so fast.

   A feeling of circumstance follows an emotional memory that is not about any individual substance or event, but about the condition in which that substance or event appeared or occurred and is general applying to variety of substances and moves. Therefore, a feeling of circumstance is an end product for a pathway in outer layer which can barrow different substances or behavior memories from the middle brain. This is clearly found in dreams, which a released stress flow from an overstrained location recalls alternative symbols which never happened in waking time. The hippocampus and the amygdale are so connected to release overstrain energies through hypothalamus structures to exhaust by expressing.

5) According to the above discussions, the type of stress flow through the outer layer is dispersive, ending in optional outputs; the type of stress flow through the middle brain ends in two options; and the type of stress flow through the inner layer is a not an optional ending—there is only one ending for these—going to motor-neurons for either expression or action. The category of stress flows can be explained according the structural neuron types. Multipolar neurons are the most common neuron type in the brain and more than 99% of human neurons fall under this category. Bipolar neurons are spindle-shaped. Unipolar neurons like sensory neurons which have only a single fiber. Illustration below.[15]

---

15 Adopted from http://www.tutorvista.com/content/biology/biology-ii/control-and-coordination/units-nervous-system.php

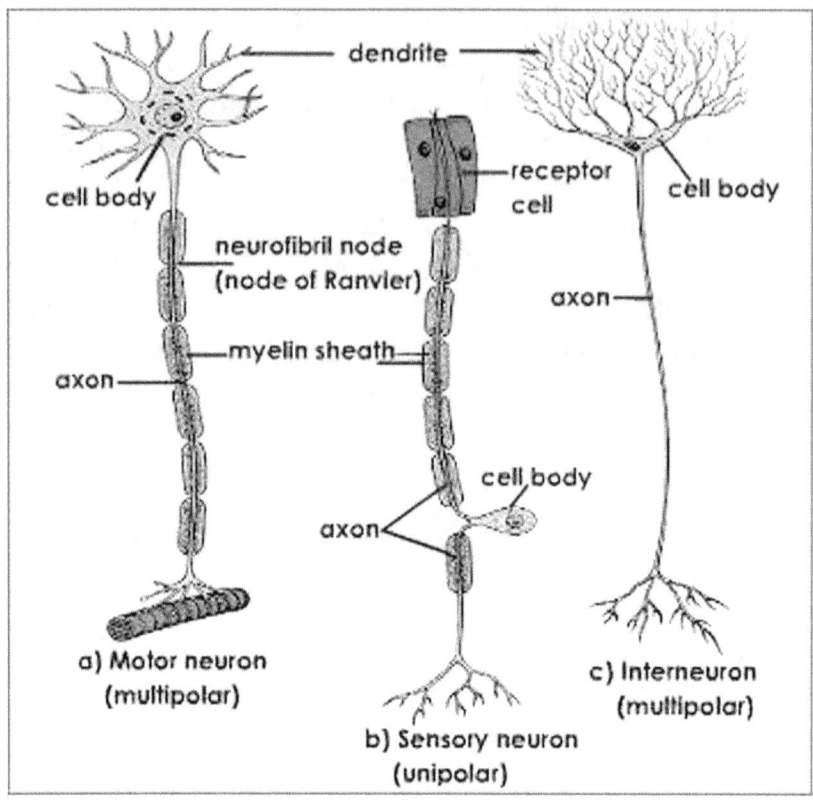

6) Generally, pathways can be distinguished by the types of neurons found in their starting points, boundaries and terminations.

7) Furthermore, the type of functional neurons can guide one to the type of mentations ending.

8) Because the brain is generally a media for the appearing and disappearing of the energy fractals and fractals are self-generated configurations, the general structure of the substrates in any part of the brain should be highly similar to their building components of neurons type and even

their sub-molecular type like genes. There are different neurons, in size and shape. Illustration below[16].

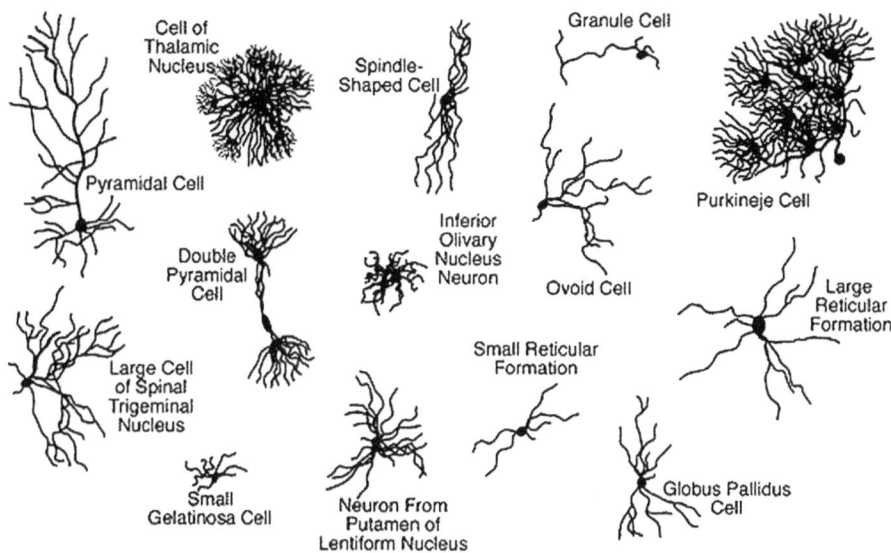

It is to be investigated if neurons like thalamic nucleus, interior olivary nucleus and purineje cells, which are found in the middle brain, store memories because they are so concentrated and hard that deformation happens to them only in sudden reconfigurations or consolidations, while neurons like Pyramidal cells, reticular formations and Granule cells make a good bed of substrates for easy energy flowing in the outer brain.

To find the location for each type, one can refer to the self-similarity in the structure of the area. For example, pyramidal and spindle locations are expressed as follows:

---

16 Adopted from "Neurons, Synapses, Action Potentials, and Neurotransmission," Robert Stufflebeam, http://www.mind.ilstu.edu/curriculum/neurons_intro/neurons_intro.php

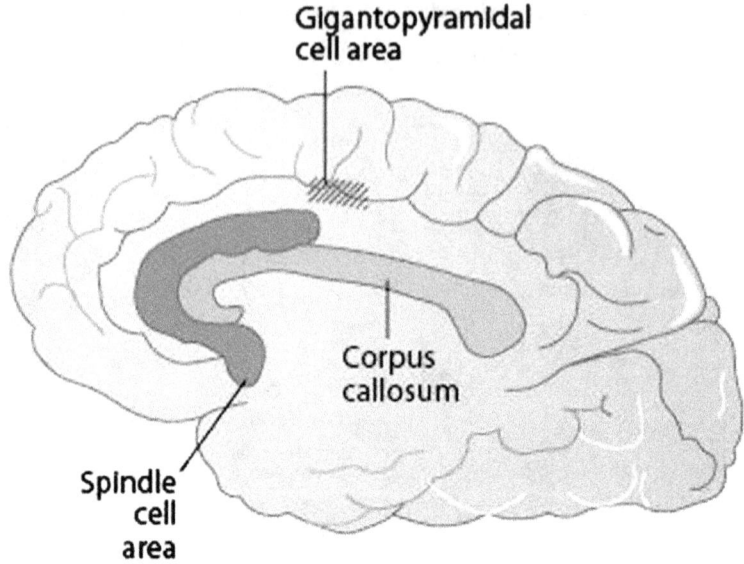

Above: Pyramidal cell area & Spindle area[17]

The cell types in the basolateral and cortical nuclear groups are similar to each other. Most of the neurons in both groups are termed pyramidal cells because they resemble the pyramidal neurons in the cerebral cortex. They have large pyramidal-shaped cell bodies and dendrites that exhibit a dense covering of dendritic spines. The pyramidal cells are the main projection neurons of these nuclear groups (i.e., their axons project out of the amygdala and allow the amygdala to activate other brain regions.

Neurons in the lateral portions of the centromedial amygdala resemble the medium-sized spiny neurons of the adjacent caudate and putamen. Most of the neurons in the centromedial nuclear group contain neuropeptides, GABA, or both. Neurons in the more rostral parts of the extended amygdala (e.g., the bed nucleus

---

17 Adopted from http://www.answers.com/topic/spindle-cell

of the stria terminalis) are similar to the cell types found in the central and medial amygdalar nuclei. The thalamus is best known as the final relay station to distill sensory information into a more interpretable and manageable form for perceptual data before it is passed on to the cerebral cortex. It receives input from diverse brain areas, primarily including all the <u>senses,</u> aside from olfaction. It is also responsible for regulating motor control and it modulates arousal mechanisms, maintains alertness, and directs attention to sensory events.

The thalamus consists of 3 circuits: the <u>specific nucleii</u>, the <u>reticular formation</u>, and the <u>intralaminar circuit</u>. The <u>specific nucleii</u> are responsible for scanning the cerebral cortex and determining active brain regions—those <u>firing at around 40Hz</u>—then relaying this information to the rest of the thalamus. <u>The reticular formation</u> is constantly making intelligent guesses as to what sensory object is generating these activation patterns. <u>The intralaminar circuit</u> compares these pattern guesses with similar patterns stored in memory. All these circuits cooperate to produce a coherent framework for the interpretation of incoming sensory data.

The hippocampus converts and directs incoming information into its appropriate memory banks. Thus, it is involved in learning through day-to-day experiences and deliberate study. Repeated use of its nerve networks enhances memory storage from short to long term, in addition to strengthening its recall. It seems to decide which memories are stored by attaching an "emotion marker" to some events for easy recall. This system has extensive influence on human behavior because most behavioral components, such as value judgments, are connected to the hypothalamus.

The hippocampus also specializes in processing sets of stimuli (not individual stimuli); or, in other words, the context of a situation. The context associated with a traumatic event can provoke anxiety because of the close connections between the hippocampus and the amygdala."

"The amygdala lets us react so rapidly to the presence of a danger, that only afterward we realize what frightened us. How? It connects events with emotions, and helps arouse survival feelings such as fear, pity, anger, or outrage. All information captured by the senses is routed first to the thalamus. It then sends each message on to the appropriate sensory cortex (visual cortex, auditory cortex, etc.), which evaluates it and assigns it a meaning. If this meaning is threatening, then the amygdala is informed and produces the appropriate emotional responses. However, a part of the message received by the thalamus is transferred directly to the amygdala, without even passing through the cortex! This shorter and much faster route explains the rapid reaction of our natural alarm system[18]

There are some vague terms in the above texts like determine, make a guess, interpret or emotion marker, which are not defined in physical terms. What is needed is to use materialistic means to translate them so that they are in line with our methodology. The following rule provides an explanation tool, accordingly.

The attraction rule: always, an established memory would be judged by a more temporary one. This means that a working memory would be judged by short, long memories; or, if a short memory would be excited, it would be judged by a long memory, no matter if it is a substance, event or circumstance memory. The reason for this is that a more established memory has a more

---

18 "retrieved from: http://www.education.com/reference/article/amygdala/.

stable configuration structure, and naturally has higher inertia energy to change. In previous volumes of this book, the concept that more stable memories work as an attractor for a resent flow of stress in the outer layer was discussed indicating the outer layer is an easy flowing stream bed for energy transfer. Having this and reminding a pathway is not aware of what is happening through it, but directs a stress flow based on streamlines that are attracted by more stable structures in other locations, gives an overview of pure materialistic brain activity.

What gives awareness about the stress flow and its endings is the emotion marks occurred during the stress transfer. Each concluding circumstance from the interactions through the pathway recalls the relevant emotion from the amygdale. The emotion works as an attractor recalling some other similar circumstances; when each circumstance memory opens for sequence of substances and moves which made that circumstance. The individual recalled substances and moves will be generalized in the outer layer. The pathway continues with recalling other similar circumstances and calls more sensory inputs to enrich the generalization and ends with emotion or awareness decline. Illustrated below[19].

---

19 Adopted from http://www.c2cinternet.eu/gpage44.html

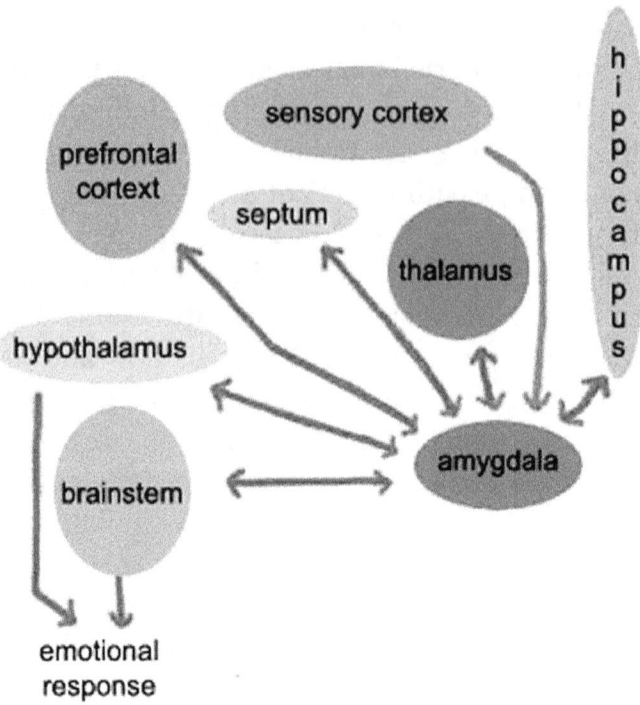

The unimodal cortex conveys the *object*; the polymodal cortex carries the *concept of the pathway* which is being developed by recalling components of circumstances with the same emotion as the *context*. All these are regulated by the governing awareness of the excited related emotion site in the amygdale. All conveyances occur through substrates made of similar neuron types, either locally connected, or transferring energy by resonance. Accordingly, the starting point in the pathway is the sensory thalamus, which is supposed to modulate and to give a definite configuration to the stress flow, which, in turn, is released from the strained substrates to waiting substrates, according to the substrates' stress-strain relationship. The excited configuration in the amygdale gives an awareness of the object and the brain activity concept, context and characteristic to be transferred to the hypothalamus, where it is judged circumstantially as agreeable or not agreeable to the preservation and self-advancing needs and requirements. Without this process in place, the result is a confused feeling of anxiety and non-determination in expression or action. Illustrated below[20].

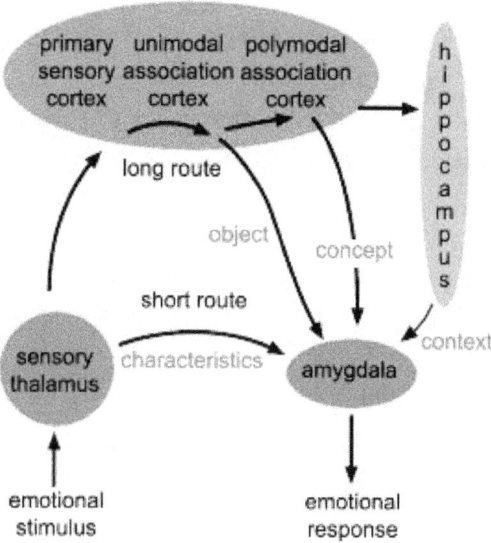

The primary and automatic bodily responses involved in emotions such as fear are controlled mainly by the outputs of the amygdala to the nuclei of the sympathetic nervous system in the brainstem and to the hypothalamus, which itself controls the hormonal secretions of the pituitary gland. The outputs of the amygdala provide a good idea of what is necessary for the experience of an emotion such as fear. The connections from the amygdala to the cortex can influence attention to, and perception and memory of, dangerous situations. The amygdala can also influence the cortex indirectly, through its connections to the attention system in the brainstem. Other parts of the brainstem trigger the cascade of physiological reactions associated with fear that send feedback to the brain.

---

20 Adopted from http://www.c2cinternet.eu/gpage44.html

The connections from the prefrontal cortex to the amygdala involve secondary emotions and enable us to exercise a certain conscious control over those emotions and thus anxiety. However, the cortex can create anxiety by allowing us to imagine the failure of a given scenario or even the presence of dangers that do not actually exist. Most of the neural pathways that enter the amygdala are paired with other pathways that exit it. The gratifying or aversive nature of a stimulus is associated by connections of this pathway to the nucleus accumbens[21].

Transferring-Storing Energy and Streambed Configuration- Reconfiguration

Synapse configurations that are involved in the stream of moving energies temporarily store the overstrain forces during the waking hours. The temporarily stored energies will be released either by some fine reconfigurations, making a memory of related inputs, or distributed in the layer as a local energy uniformity around the overstrained point. Moving energies are carried by momentums.

The brain shows different properties in relation to energy transfers:

1) Its components work as an energy capacitor and temporarily confine, or permanently absorb, the energy.

2) Its curved circuits work as coils that synchronize firings upstream and induce resistance forces, as well as imposing forces, to change the circuit curvatures.

3) Its elasticity property, on different scales, works as a spring, absorbing inertia forces by substrates' synapses configuration change, to balance the inertia forces.

---

21 Adopted from http://www.c2cinternet.eu/gpage44.html

All of these properties are combined by the key properties of viscoelasticity and nonlinearity. Because of this combination, the brain's positions finely waves to balance these energy transfers and the energy storing properties in the affected areas.

Momentum transfer, the dissipation of energy achieved by confining the configuration energy and the reconfiguration of synapses curvatures all change firings gate sizes and distributions. To find an order in gates' firings' oscillation and create a clear map of moving-storing-recalling energy distribution, it needs a high driving force input, initiated either from environmental or internal sources. Any variation in the driving force changes the regulated oscillation map and geometry functional outcomes.

The synapse configuration ordering parameters are oscillation frequency and synchrony. The configuration is a system that is far from a state of equilibrium and that dissipates energy by absorbing it in its fine structures. Therefore, the brain is a self-organizing system creating dynamic energy fractals, as well as developing the growth and functioning. Ordered patterns will appear when driving force and energy absorption are of the same magnitude orders. However, during waking periods, the layer elasticity will be degraded and, consequently, stress absorption will be reduced and temporary created maps will be distorted.

Mathematically, the driving force (F) equals changes of momentum per time ($d\tau/dt$); the driving parameter is the momentum transfer velocity (υ); energy flow is the product of driving force and driving parameter ($\upsilon * d\tau /dt$) and resistance against energy transfer is the reverse of changes in drive parameter per driving force ($\Delta\upsilon/ \Delta\tau)*dt$.

Different oscillation maps can appear at the same time, combining or flowing in parallel. Therefore: 1) maps can follow intervallic oscillations, or 2) maps can combine into a complex if the number and directions of progress fulfill the requirements for making a complex.

Position Oscillation in Substrates

If the inertial per shear force ratio for a substrate changes, the substrate tends to display a kinetic character, which could be a regular motion of associated energy packets or a disturbed behavior. The inertial per momentum ratio is a variable, while inertia plus momentum for the travelling energy through a substrate is constant according to the conservation law. Accordingly, the inertia magnitude changes from a minimum to a maximum and the ratio will fluctuate.

Changes of the ratio indicate variations of substrate's amplitude dispersal between a minimum and a maximum limit as tissues will permit in the location. The fluctuation between the minimum and maximum levels indicates an oscillation in the synchronized firings or energy packet transferring in the substrate. The measured frequencies of a substrate indicate its structure's natural frequency, which is a spatial combination of their configurations within the substrate.

The Brain Develops a Sense of Circumstances and Possibilities into Countable Logical Analysis

- Sensory inputs as facts and events and internal initiators such as felt circumstances would map a pattern in brain, while will go under functional changes according to the relevant attractors like locations with a saved memory attached to similar circumstance feelings. The final product of the interaction is either an output product of an expression or action. Maps, according to the type of neurons and their special structuring in the networks in a location, are made of different configuration structures in the outer layer and the frontal lobe.

- The feeling of *circumstances* that any individual exposed to the same environment finds would be different in interpretation based on the internal attractors. Due to varying attractors, *alternative pathways* would be developed in the outer layer, mapping the energy flows. Concurrent energy flows interact together and reasoning depends on the numbers of the concurrent, inputs and complexes of configurations they may form. This process of interaction targeting the needs satisfaction is a random optimization, conducting the sequence of inputs-outputs orders. By this process, a state change from a chaotic to a laminar for the energy flow; or a transformation from fuzziness to coherence in cognition occurs, which is known as logical reasoning.

- As free energy travels from region to region, the visco-elastic property of a substrate displays a flexible range of strain responding to stresses. However, the free energy transfer ends when the transferred momentum is too weak in absorption and too strong to dispatch. The ending substrate would be tensed for an intermediate strain-mark. According to the pathway's functionality, the strain marks would be temporary saved or permanently consolidated. For momentum transfer strengths that fall between these limits, the output signal contains a degree of fuzziness.

- The circumstance feeling is accompanied by tension and strain; and if the inducing strains goes beyond a limit, it introduces disorderly behaviors to release excess energy. "A depolarization of only 3 millivolts causes synaptic depression and of 10 millivolts causes synaptic potentiation[22]."

- Saved memory of circumstances reinforces functional substrates in the pathway strain-mark. The strain-mark location will work as an attractor in future inputs which induce the same circumstance feeling and that becomes as

---

22 Linden, *The Accidental Mind*

a specialty for the location. For example, different memories of sounds in the auditory cortex and memory of scenes in the visual cortex are specialties for the locations which resonate with the same frequency of inputs. A complex of the specialized substrates as a pathway for a complex of related inputs conducts more advance cognition processes.

- Saved memories, *by frequent recalling*, reinforce the specialty of the substrates by increase in number of recalls and excitation and become a functional memory. Functional memory is an integration of similar, or a family of memories which refresh the same circumstance feelings. When the recalls do not happen frequently, configuration may become deformed and specialty would be degraded.

- Working memories are short life memories attached to the pathways with the confined strains. They reveal the trace of resent interaction between inputs and functional memories.

- The difference between a memory substrate in the middle brain and a functional substrate in the outer brain is that a solid memory substrate needs a higher inertia to change its synapses configuration, and so acts as an attractor, while functional substrates are easy flowing beds for energy transfers and need less inertia to find a continuous change in the configuration. In other words, memory substrates are made of a very local and concentrated spatial network of synapses and are localized; when, functional substrates are made of very fluent and flexible configurations within the limit of elasticity.

- In any case "... the synapses of the brain are not static. They grow, shrink, morph, die off and are newly born, and this structural dynamism is likely to be central to memory storage[23]."

Circumstances and Substances-Events

Emotions are the language of inputs. They tell us with such an inputs, what *circumstance* is at the present, or probably is going to be around us, in respect to survival and the satisfaction of needs and desires. Thoughts express the circumstance in respect to input sensations, by imagination of the short excited pathway. Daily stress releases from the excited pathway are disposed by *acts* and *expressions*, which are achieved through words or behaviors.

Focused thinking describes the possibilities in the present and future through the analyzing, synthesizing and generalizing of initial parts of the pathway, where inputs are entering and, consequently, by creating imaginations of alternative pathways in a combination, the resulting feeling and thoughts of circumstance will be complicated. By challenging of the complex with new inputs, feeling and thoughts errors will be reduced in total. The resulting strains through the pathway will be released by expressions of feelings, thoughts and thinking, or by performing a behavior.

In sleep, the remained strains will be released entirely by expressions (except some behaviors which are known as disorders) of feelings and thoughts refreshing the severe felt circumstances against needs. Feelings of circumstances, which were very strong in term of strain consequences will be reprocessed in regeneration periods to either be released (in REM periods) or be consolidated (in NREM periods) by new configuration of some synapses. Temporarily confined strains produce an unstable structure of synapses configuration in the related substrates which is needed to be released to find a stable and *stress-free configuration*. The

---

23 Ibid

releasing phenomena in a distorted pathway which are initiated by most-strained locations provide a flow of stresses, saving or recalling in the middle brain substrates as attractors. The new established substrates convey the same feelings anytime will be excited in the future.

Forcing Instinct, Dictating Thought and Supervising Thinking

The simultaneous running of a number of mentations, beyond a threshold, creates a conscious complex. Man talks to himself, shows sympathy, criticize, judges, evaluate and optimize actions based on just such a conscious complex, which is regulated in the frontal lobe.

Species of lower complexity are governed either by 1) instincts, which direct their reaction to excited pathways or by 2) free or automatic thoughts, which are more complex urging them how to express or behave. 3) Thinking or planned thoughts extrapolate the present thoughts to cover some more inputs in future. Thinking brings options of inputs together simultaneously, although the process of choosing or weighting them is ordered by a complex of all those components interactions and by the initial condition and efficiency of interactions. However, the will as a result of thinking is also limited by what an individual's attractors (needs and desires) will permit. Complex attractors have a strong effect in directing pathways and control the ratio of basic attractors' involvement in forming of a pathway. The product of the formed pathway is a final product of a choice, a decision or an action, or intermediate products of maybes and if-statements to be continued for coming final products.

The pathways for thinking activities are extended to the frontal lobe, with its special configuration of nerves, synapses and firing gates distributions, as well as its dense connections which make such a combination possible compared to other areas.

Reconfigurations Become Limited and Regeneration Deteriorates with Age

Deep sleep and synapse growth hormones have an intensive relationship. According to research, the reducing of REM periods after middle age is significant. Although it is not known that less deep sleep causes reduction in growth hormones or that reduction in growth hormones results in less deep sleep, it seems that the inability to geometrically balance the brain within the head skeleton by fine curls is the main condition that causes the regeneration process in both REM and NREM periods to become imperfect and inefficient. The evidence for this is the fact that deep sleep begins in young adulthood, when the skeleton becomes solid and fixed.[24]

Emotions, Configuration Energy Transfer and Pathways

In this section a comparison between kinematic model and present theories for any brain activity from initiation to termination will be undertaken. In the explanation of the present theories that follows, all the contents and images are summarized and taken from the other sources, as referred to.

The Cannon-Bard Theory: When a person faces an event that somehow affects him or her, the nervous impulse travels straight to the thalamus, where the message divides into two parts:

1. One part goes to the cortex to originate subjective experiences like fear, rage, sadness, joy, etc.

2. The other part goes to the hypothalamus to determine the peripheral neurovegetative changes (symptoms).

---

24 "Growth Hormones and the Effect on Sleep", Art Hister, MD, retrieved on November 24, 2010.

According to this theory, physiological reactions and emotional experiences occur simultaneously.

Kinematic model: The transfers in the hippocampus and amygdale (in the middle brain), as well as the frontal lobe and inner brain, are simultaneously resonating (through projections or copying) transfers in the outer layer. All those transfers are referred to as attractors' dynamics. There is not any major discrepancy between. Illustration below[25].

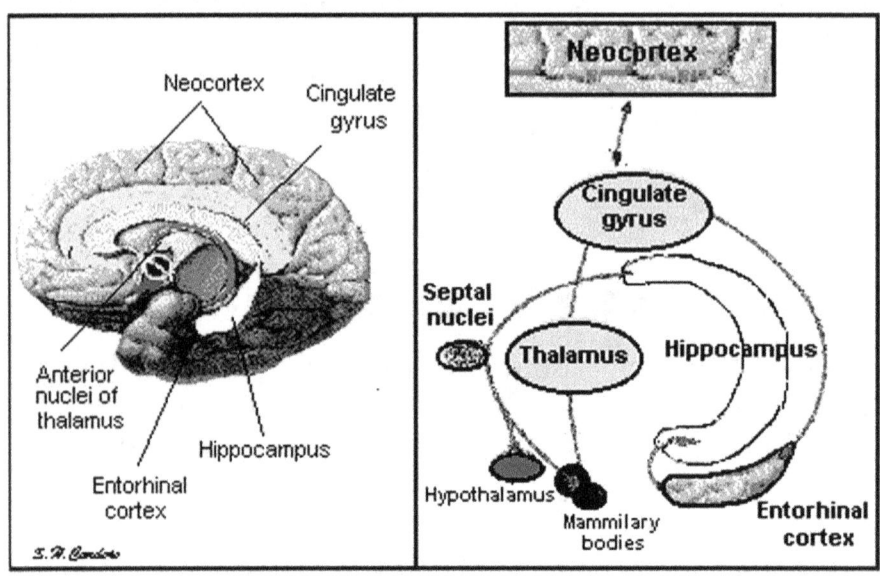

Paul McLean proposal:

A summary of the Paul McLean pathway description for brain activities is as follows:

1. The cingulate gyrus projects to the hippocampus, and

2. The hippocampus projects to the hypothalamus by way of the bundle of axons called the fornix.

---

25 From Wikipedia; The Limbic System

3. Hypothalamic impulses reach the cortex via relays in the anterior thalamic nuclei. The essential error of the Cannon-Bard theory was to consider the existence of an initial "center" for emotions (the thalamus).

4. Emotional expression was thought to be governed by the hypothalamus.

Kinematic model: An emotion is an overstrain location of a pathway over the outer brain, and will be distributed to equalize the layer. The distribution transfer has a resonating parallel transfer over other regions, including the hippocampus and amygdale (in the middle brain), as well as the frontal lobe and the inner brain, saving or recalling attractors (i.e. characters, behaviors and emotions, as well as reasoning and reflexes).

James Papez explanation:

Emotion is not a function of any specific brain center but of a circuit that involves four basic structures, which are interconnected through several nervous bundles: the hypothalamus with its mamillary bodies, the anterior thalamic nucleus, the cingulate gyrus and the hippocampus.

This circuit (Papez circuit), acting in a harmonic fashion, is responsible for the central functions of emotion (affect), as well as for its peripheral expressions (symptoms): the orbitofrontal and medialfrontal cortices (prefrontal area), the parahippocampal gyrus and important subcortical groupings like the amygdala, the medial thalamic nucleus, the septal area, prosencephalic basal nuclei (the most anterior area of the brain) and a few brainstem formations. Papez believed that the experience of emotion was primarily determined by the cingulate cortex and, secondly, by other cortical areas.

All these structures interconnect intensively and none of them is solely responsible for any specific emotional state. However, some contribute more than others to specific kinds of emotion. Paul McLean added new structures to the circuit, including the orbitofrontal and medialfrontal cortices (prefrontal area), the parahippocampal gyrus and important subcortical groupings like the amygdala, the medial thalamic nucleus, the septal area, prosencephalic basal nuclei and a few brainstem formations.

Kinematic model: Emotion is not a function of any specific brain center but a simultaneous excitation of resonated locations in the hippocampus and amygdale (in the middle brain), as well as frontal lobe and inner brain areas, through the process of overstrained locations relaxing in the outer brain and recalling saved information in the excited areas. In the kinematic model, the detailed locations which are mentioned, as the Paul McLean circuit, are also located in the brain layers.

Let us examine the kinematic views provided by the model in a more detailed way:

Stressful inputs, which enter convergent in time and space, may result in strained location in their related pathway. Inputs that are convergent in time are entered through sensory channels should happen in the same time (practically in a small enough time frame).

Inputs that are convergent in space are processed through one unique pathway in the brain and saved in the same brain location. Any pathway, in any layer, can initiate an induced pathway in other layers through resonating and a copying process that requires the initiating pathway to attract, recall or save configurations in one layer from, or to, the other layer. The saved configuration plans in different layers will consolidate periodically, in accordance with the existing curvatures in that layer. Furthermore, the copying of information will be carried out differently in each layer, according to a given layer's structural fibers configuration. Therefore, the entry of a set of information from the environment into the input zone kinetically transfers, as an energy flow, into the outer layer; meanwhile, the middle layer either fits an existing configuration as a known character or behavior or fixes a new configuration as a new static memory (like shapes, names and numbers), a dynamic memory as an new event or move (in hippocampus). Similarly, the parallel resonating flow recalls and presents an emotion type (in amygdale). The energy transfer will be saved as a modifying function in the outer layer, as a memory in the hippocampus, as an emotion type in the amygdale and as a way of reasoning in the frontal lobe.

In a similar process, during the regeneration of layer elasticity property in sleep, the releasing of the confined energy from a strained location recalls a plane of memories from the hippocampus and excites an associated emotion type in the amygdale. The memory plan in the hippocampus consists of a set of characters and behaviors which have been saved in different input entries and fit in the same topology of the local configuration attached with a specific emotion type.

This model is the basis of a product called Dream Pack, which is under development. According to Mariam Joel from the University of Amsterdam:

- "Convergence in time is when the stress hormones are released during or immediately after the event to be remembered."

- "Convergence in space occurs when the hormones released in response to stress activate the same neuronal circuits as does the information to be processed and stored[26]."

Accordingly, memories of collections of different characters and behaviors, which have been converged for time and space, can be saved in the same location in the hippocampus, one that is associated with the same emotion type, although the boundaries of different collections are different. This is the reason that a

26 Mathias V. Schmidt and Lars Schwabe. "Splintered by Stress," *Scientific American Mind*. Sept 2011.

strained location, when released in a dream, brings up a confused mixture of characters and behaviors which initially seems meaningless. However, the common point between them is a shared associated emotion type. In other words, a created collection of characters and behaviors from different saved memories with the same emotional theme make a dream. The shapes or characters and moves and behaviors in different collection scenes are symbolically or metaphorically the same, if they initiate the same emotion type.

Kinematics Model and questions

By now some of the following questions have been answered by the model, but some still need to be answered. The answered questions are:

- How are the active brain locations in waking time and in sleep periods (pathways of mentations and the over-strained locations) connected?

- How are highly emotional mentations during waking times linked to the emotions felt in dreams?

- Why can sleeplessness kill a person?

- How is a mentation a self-organization of energy packets?

The non-answered questions follow, as well as an attempted answer for some of them:

- What are the main factors in the strain releasing process: the duration of imposing stressful inputs, the duration of sleep stages or REM and NREM details?

- How can the correlation between the imposed daily events and dreams be formulated?

- Future designed experiments should confirm evidence of the back and forth sliding of synapses, axons or fibers inside their isolating wraps or surrounding non-nerve tissues .

- What are internal initiator's and attractor's role in this correlation? Is it determined through the body metabolism activities, unbalancing between satisfied needs and instincts or genetic influences or growth metabolism?

- Is "a day of absorbing stresses and a night of regeneration" a complete cycle or is it not a complete cycle, in which case it would leave remains or a balance of strains to be processed in the following nights?

- Are strains due only to one stressful input or a mix of them? If a mix, how much of the related building memories are reliable?

- Why are stressful days followed by sleepless nights? Does this mean that highly energetic internal stimulus prevents sensory channels from closing and asks the disordered front lobe to work, even in a disordered and inefficient state? Or, does unbalance in the input area hurt tissues?

- Is brain convolution growth or unfolding an influencing factor on people's number of sleep hours?

- Sleeping more results in more sleeping or in other words regeneration period is stressful itself. Does it mean the regeneration process is possibly not convergent and that layer ground energy is fluctuating around in a REM state? Are there NREM stages for extra hours of sleep or does reconfiguration find a cyclic equilibrium level to fluctuate around that?

- Disorders in sleep normally accompany disorders like in eating, worries, depression, and likely addiction. Is this due to layer geometry disturbance and a lack of geometric balance in a location?

- Firings orbits and fractals are static structures. How can a dynamic behavior be related to energy wakes consisting of firings orbits or fractals?

- Is the reconfiguration of synapses (which is defined as the consolidation of memories in NREM and is assumed as a fact indicated by oscillation shape picks) a regular fact? How it can be proven?

- Is there any evidence for the physical vibration of positions in brain layers during the transfer of energy packets? What is a realistic estimate of the size of the substrates amplitudes?

- Which method is most appropriate to measure and demonstrate energy transfer through pathways?

CHAPTER III

Self-Inducing Currents and Sensory Inputs

In any direction of synapse connections, the neuron fibers positions are far from straight lines. Firings' synchronization development and free energy transfer along a route of synapses and neuron fibers induces an electromagnetic field. The electromagnetic field induces a force on the fibers according to the rules of electromagnetic fields. The created field induces a flow of stress along the nerve strings within an effective field radius, and this is a self-induced current. The natural induction in connected nerve fibers is similar to the inductions applied by MFR or mFNMI methods to brain fibers, which accordingly detect magnetic oscillations in the brain. A nerve is normally made of several nerve cell fibers joined with connective tissues. The nerves are insulated by a wrapping sheath. The external coil in these devices is to be considered as curvatures of the nerve strings. Illustration below[27]

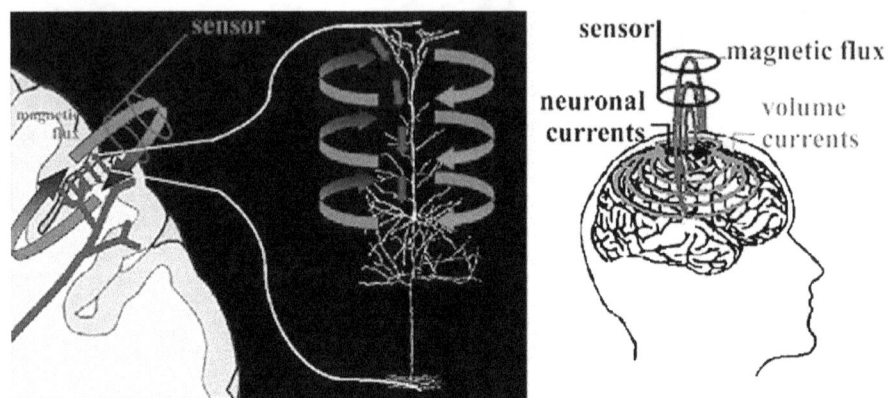

The Way that Self-Organizing Patterns Grow

The self-organization of stress stream patterns proceeds as follows. Synchronizing will be proceeded by the induction of an electromagnetic field, which is created by each substrate of nerve fiber networks into the next substrate, thus synchronizing the next substrate firings of its neuron fibers network. Accordingly, self-organization will be developing up to the point that similar patterns of orbits and fractals would be created. The appearance of the overall pattern will be accelerated by micro-scale catalytic properties of genes in the molecular sites of substrates as well as the macro-scale elasticity properties of the substrates.

---

27 Adopted from Magnetic Stimulation of Neural Tissue http://www.bem.fi/book/22/22.html

Application of the Reptation Theory to Fibers and Their Wrappings

The reptating of fibers, axons or synapses within their isolating wraps results in the minute back and forth sliding of an entangled string fiber in a network of fibers within a substrate that is under the induced electromagnetic forces of other substrates.

Any stress force on a string causes such a forth slide, which allows it to balance itself. Also, any sting that experiences a forth slide, after the stress release, will return back because of the elasticity property of the string. However, the whole network remains under an oscillation of back and forth movements if the media that is made by them is in a chaotic condition. Consequently, any back and forth sliding carries an entangled packet of energy. A full cycle of the back and forth sliding is a spring-like movement limited to the viscoelasticity property of the tissue and therefore the sliding transmits a flow of stresses as energy packets.

Future designed experiments should confirm evidence of the back and forth sliding of synapses, axons or fibers inside their isolating wraps or surrounding non-nerve tissues.

Accumulations of New Curls and Folds

Lipofuscin, an aging component (called wear and tear), is arranged around the nucleus of the neurons, making new convolutions and folds. This component is not responsible for fixing synapse curvature, it is an indicator for the tissue curvatures. Therefore any new consolidated memories in the regeneration periods can be traced using this chemical. Because the brain surface is free of bounds with other tissues, the configuration of new curls and curvatures finally appear as new folds over the course of decades, and the general pattern of folds should have a relation with an individual behavior as his or her characteristics.

Conclusion: The Need to Apply the Kinematic Approach to Brain Function Modeling

- The kinematic approach to modeling brain functions provides a macro-scale perspective in change of forces, energy transfers and higher level changes, rather than a micro-scale perspective, as is seen at a molecular level. It is a new kind of approach and can countercheck neuroscience and psychology findings.

- There are many proven theories to take as a basis for building a kinematic model. However, additional assumptions and claims are required to complete the model sufficiently and to support energy studies in firing energy packet sizes.

- In parallel to the kinematic study of brain behaviors, the applied assumptions and claims are to be modified or substituted by more correct ones, keeping contingencies within the model.

- The need for kinematic study is due to the fact that studies on the micro scale level are not able to answer all the questions about brain behaviors. The model should be tailored in a trial way in order to answer the global questions concerning memories, moods, and the nature of sleep and its stages.

Glossary

Attractor: the substrates anywhere in the brain which influence the direction of the main stream of the free energy flow pathway over the outer layer.

Chaos theory: Chaos theory studies the behavior of a dynamic system that is highly sensitive to the initial condition and inputs.

Energy Wake: a temporary stagnant energy staying in a location; is removed when location goes to a rest position.

Euler Equations: represents conservation of mass (continuity), momentum, and energy for an ideal media.

Firing fractals: it is assumed that because synapses grow in fractal shapes, the firing gates distribution in a location are arranged in a fractal shape and consequently the synchronized firings in the location produce a fractal of firings

Fluid Dynamics: the natural science of fluids in motion. Although the fluid is defined as a gas or liquid, which is the media for the energy transfer, here the brain is the media which a free energy travels through it trying to reconfigure the synapses' configurations.

Kinematics: Kinematics describes the motion of synapse positions and systems (groups of synapse positions) without consideration of the electromagnetic forces, which are induced by synchronized firings through a pathway and cause the motion.

Mentation: a continuous activity of the brain in the shape of free energy transfers which form an introversion or extroversion pulse to communicate or command for an action.

Natural Frequency: the frequency at which a substrate vibrates when the substrate is considered as a set of individual fibers with free vibration in the fiber network of the substrate.

REM/NREM: different stages that occur during sleep.

Reynolds' Number and Index: a non-dimensional number that determines which type of flow is present and it is a ratio of inertia forces to viscous forces of free energy through a pathway made of communicating synapses.

Stress-Strain Relationship: is the relationship between a strain caused by an input stress on different locations of a pathway of connected synapses.

Synchronized Firings: the firings happen in the same time and normally should be locked in the same phase.

 Viscoeslasticity/Viscoelastic Properties: the property of materials that exhibit both viscous and elastic characteristics when undergoing strains.

Index